Polymer Rheology

Polymer Rheology

Lawrence E. Nielsen
Monsanto Company
St. Louis, Missouri

MARCEL DEKKER, INC. New York and Basel

Library of Congress Cataloging in Publication Data

Nielsen, Lawrence E
   Polymer rheology.

   Includes bibliographies and indexes.
   1. Polymers and polymerization. 2. Rheology.
I. Title.
QD381.8.N53   668   77-24187
ISBN 0-8247-6657-1

MARCEL DEKKER, INC.

270 Madison Avenue, New York, New York 10016

Current printing (last digit):
10 9 8 7 6 5 4 3 2 1

PRINTED IN THE UNITED STATES OF AMERICA

To
Deanne,
Bryan,
and
Kimberly

CONTENTS

Preface ix

Chapter 1. INTRODUCTION TO POLYMER RHEOLOGY 1

    I. Rheology 1

    II. Units 3

    III. Measurement of Viscosity 3

    IV. Normal Stresses 8

    V. General References 9

Chapter 2. INSTRUMENTS 11

    I. Introduction 11

    II. Capillary Viscometers 12

    III. Coaxial Cylinder Viscometers 18

    IV. Cone and Plate Rheometers 20

    V. Parallel Plate Viscometers 22

    VI. Tensile or Extensional Viscometry 23

    VII. Dynamic or Oscillatory Rheometers 24

    VIII. References 27

Chapter 3. EFFECTS OF TEMPERATURE AND PRESSURE 31

    I. Temperature Dependence of Viscosity 31

    II. Master Curves for Temperature Dependence 36

    III. Effect of Pressure on Viscosity 42

    IV. References 44

Chapter 4. EFFECTS OF RATE OF SHEAR ON POLYMER RHEOLOGY 47

    I. General Non-Newtonian Behavior of Polymer Melts 47

    II. Role of Entanglements 48

III.   Theories Describing Shear Rate   51
        Dependence of Viscosity

IV.   Dynamic Properties   52

V.   Rheology of Emulsions, Polyblends,   54
     and Block Polymers

     A.   Emulsions   56

     B.   Polyblends, Block, and Graft   61
        Polymers

VI.   References   65

Chapter 5. EFFECTS OF MOLECULAR WEIGHT AND   69
         STRUCTURE

I.   Molecular Weight Dependence of   69
     Viscosity

II.   Effects of Distribution of Molecular   72
     Weights on Viscosity

III.   Dependence of Dynamic Mechanical   75
        Properties on Molecular Weight

IV.   Effects of Structure on Polymer   80
     Rheology

     A.   Branching   80

     B.   Other Structural Factors   82

V.   References   83

Chapter 6. EFFECTS OF SOLVENTS, PLASTICIZERS, 87
         AND LUBRICANTS

I.   Rheology of Polymers Containing   87
     Solvents and Plasticizers

     A.   Introduction   87

     B.   Viscosity as a Function of   89
        Concentration of Solvent

     C.   Temperature and Shear Rate   94
        Dependence of the Viscosity of
        Polymer Solutions

     D.   Dynamic Rheological Properties of   96
        Solutions

II.   Lubricants   99

III.   References   101

Chapter 7. NORMAL STRESSES AND DIE SWELL   105

    I. Normal Stresses   105

   II. Die Swell   111

 III. References   116

Chapter 8. EXTENSIONAL FLOW AND MELT FRACTURE   121
           PHENOMENA

    I. Extensional or Elongational Flow   121

   II. Melt Fracture and Flow Instability   125

 III. References   129

Chapter 9. SUSPENSIONS, LATICES, AND PLASTISOLS   133

    I. Rigid Fillers - Newtonian Behavior   133

   II. Rigid Fillers - Non-Newtonian Behavior   142

 III. Rheology of Latices   150

  IV. Rheology of Plastisols   152

   V. References   154

Chapter 10. RHEOLOGY OF POWDERS AND GRANULAR   159
            MATERIALS

    I. Importance of Powder Rheology   159

   II. Instruments   160

 III. General Flow Behavior of Powders   164

  IV. Polyvinyl Chloride Plastisol Resins   173
        as an Example of Polymer Powders

   V. References   176

APPENDIX

   List of Symbols   179

AUTHOR INDEX   185
SUBJECT INDEX   201

Why is there the need for another book on rheology? The answer is that there is no book on the subject that is suitable for the majority of people who are actually working with polymers and who need to know something about rheology in order to properly do their job. What does the average scientist, engineer, or student, who is not a specialist in polymer rheology, want to know about the field? I believe this worker wants to know one or more of the following: 1. The worker or student wants to know the general behavior of polymers during flow and a phenomenological explanation of the behavior so that he develops a "feeling" for the field. 2. He wants information on the effects of the many factors which influence polymer rheology. 3. He needs to be able to make practical calculations and estimates of properties, so the required equations must be provided, but he has little interest in detailed mathematical derivations. 4. Finally, the practical rheologist or student wants to have available additional carefully selected references to the literature on any given aspect of rheology. It is the objective of this book to fulfill these needs since most books on rheology are either so elementary and superficial that they give little of practical use on the flow behavior of polymers, or they are so advanced and mathematical that only the specialist can benefit from them. This book is for the people who are actually working with plastics and rubbers and who need to know something about the flow properties of these materials. These people include research scientists, polymer processing engineers, equipment designers, fabricators and molders of plastic parts, and graduate students in polymer and materials science.

In this book, emphasis is on general principles and practical implications of all phases of polymer rheology. The treatment is at the intermediate level with a minimum use of mathematics. In general, only the final equation which can be used to make practical calculations is given. The coverage is broad and is intended to cover nearly all aspects of polymer rheology except dilute solution viscosity.

This book is unique in that it covers several topics not considered in most books on polymer rheology. In addition to instrumentation, viscosity, melt rheology, normal stresses, and other flow phenomena, additional chapters discuss the rheology of suspensions, emulsions latices, and plastisol resins. These topics are extremely important in many manufacturing and fabrication processes and in the flow behavior of composite materials, polyblends, and other two-phase systems. There also is a chapter on powder rheology. This little discussed subject is important in the transfer of polymer powders and pellets and in the operation of the first section of injection molding and extrusion machines.

The coverage of the vast literature in the field of rheology is extensive but is not intended to be complete. Rather, the references to the literature are designed to allow the reader to quickly become familiar with a few of the papers on any given topic. The reader then easily can advance his knowledge from the listed references.

I am indebted to many people who helped me in the preparation of this book. Colleagues who have offered numerous valuable suggestions after reading the original manuscript include Richard L. Ballman, Robert A. Mendelson, Eli Perry, and Murray Underwood. Mrs. Bobbie Kaplan had the formidable task of typing the manuscript. My wife, Deanne, proofread the manuscript and helped compile the indexes.

Lawrence E. Nielsen

Chapter 1

INTRODUCTION TO POLYMER RHEOLOGY

I.	RHEOLOGY	1
II.	UNITS	3
III.	MEASUREMENT OF VISCOSITY	3
IV.	NORMAL STRESSES	8
V.	GENERAL REFERENCES	9

I.   RHEOLOGY

Rheology can be defined as the science of the flow
and deformation of materials. In this work, however,
the meaning of rheology generally will be restricted to
fluid rheology. For many simple fluids, the study of
rheology involves the measurement of viscosity. For
such fluids, the viscosity depends primarily upon the
temperature and hydrostatic pressure. However, the
rheology of polymers is much more complex because polymeric fluids show nonideal behavior. In addition to
having complex shear viscosity behavior, polymeric
fluids show elastic properties, normal stress phenomena,
and prominent tensile viscosities. All these rheological properties depend upon the rate of shear, the
molecular weight and structure of the polymer, the concentration of various additives, as well as upon the
temperature.

The subject of rheology is extremely important for
polymers. In the use of polymers, it is generally the
mechanical properties which are important. However, the
mechanical behavior of an object is of little interest

if the object first can not be fabricated quickly and
cheaply. In nearly all cases, flow is involved in the
processing and fabrication of such materials in order
to make useful objects. Flow behavior is important in
injection molding, compression molding, blow molding,
calendering, cold-forming, and the spinning of fibers.
Rheology is important also in the formulation of polymeric materials in preparing them for the fabrication
process. Mill rolling and extrusion processes are
typical examples.

Rheology is involved in many other aspects of
polymer science. For example, many polymers are made
from emulsions of monomers in stirred reaction vessels.
The resulting latices flow through pipes and may end up
as a paint which is applied to a surface by some process
in which the rheological properties of the latex must be
controlled carefully. Plastisols, which are a suspension of a polymer in a liquid, are fabricated into useful objects by processes such as rotational molding.
Powdered polymers or granules must flow from bins and
must perform properly in a fabrication process such as
rotational molding and in powder coatings. The rheology
of polymer powders is important also in the first
sections of extruders and in injection molding machines
before the polymer softens to a liquid.

Rheological behavior influences the mechanical
behavior of a finished object. For example, molecular
orientation has dramatic effects on the mechanical
properties of molded objects, films, and fibers. The
kind and degree of orientation is largely determined by
the rheological behavior of the polymer and the nature
of the flow in the fabrication process.

Elasticity phenomena are a prominent feature of the
rheology of high polymers. These phenomena manifest

themselves in elastic shear moduli, normal stress
effects, and die swell in spinning and extrusion processes. The basic cause of elasticity is the orientation of molecular segments during flow. Polymer
molecular weight and its distribution also are primary
factors in all aspects of polymer melt rheology.

II. UNITS

The traditional unit of viscosity is the poise,
which has the dimensions of dyne second per cm^2 or gram
cm/sec. The viscosity of water is about 0.01 poise.
Typical polymer melts have viscosities generally of the
order of 10^3 to 10^4 poises. The presently approved
units are based on the MKS units or the SI units. The
SI units for viscosity are Pa·s. Pa is the abbreviation
for Pascal, which is Newtons per square meter (N/m^2).
The Newton in turn has the units of kg·m/s^2. To convert
poises into Pascal seconds, multiply the number of
poises by 0.100. The cgs units for elastic moduli are
dynes/cm^2, while the SI units are Pascals or Newtons/m^2.
Dynes/cm^2 can be converted to Pascals by multiplying by
0.100. The SI unit for force is the Newton, while
pressure and stresses should be expressed as Pascals.
The obsolete English units for modulus are pounds/square
inch or psi. Pounds/in^2 can be converted to Pascals by
multiplying by 6.895 x 10^3.

III. MEASUREMENT OF VISCOSITY

Ideal fluids are called Newtonian. Their viscosity
is independent of the rate of shear. The measurement
of Newtonian shear viscosity is illustrated schematically by the top half of Figure 1. A liquid is confined
between two flat plates of area A separated by a distance D. A force F is required to move the top plate

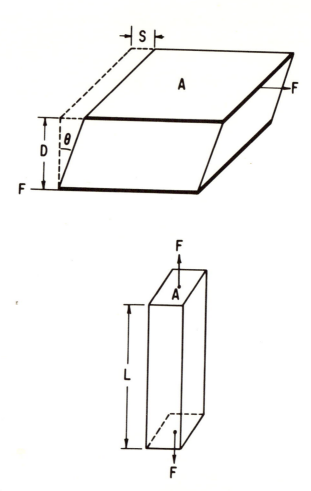

Fig. 1.   Schematic diagrams for the measurement of
          shear viscosity (top) and elongational or
          tensile viscosity (bottom).

at a constant velocity relative to the lower plate.
This force is directly proportional to the viscosity
of the liquid. The important quantities involved in

measuring shear viscosity may be defined as follows
with the help of Figure 1:

$$\text{Shear stress } \tau = \frac{\text{Shear force } F}{\text{Area } A \text{ of shear face}} \tag{1}$$

$$\text{Shear strain } \gamma = \frac{\text{Amount of shear displacement } S}{\text{Distance between shearing surfaces } D} =$$

$$\tan \theta \tag{2}$$

$$\text{Viscosity } \eta = \frac{\text{Shear stress}}{\text{Rate of shear strain}} = \frac{\tau}{d\gamma/dt} = \frac{\tau}{\dot{\gamma}} \ . \tag{3}$$

If the fluid is not Newtonian, a plot of shear
stress τ against the rate of shear $\dot{\gamma}$ is not a straight
line but a curve such as the solid line shown in
Figure 2. The liquid may be Newtonian at very low
shear rates to give a limiting viscosity η_o from the
initial slope of the τ versus $\dot{\gamma}$ curve. When the τ-$\dot{\gamma}$
curve is nonlinear, the viscosity may be defined in two
ways for any given rate of shear as illustrated in
Figure 2. The apparent viscosity η_a is the slope of the
secant line from the origin to the shear stress at the
given value of shear rate, that is,

$$\eta_a = \tau/\dot{\gamma} \ . \tag{4}$$

The slope of the line at the chosen value of $\dot{\gamma}$ is
another viscosity called the consistency η_c.

$$\eta_c = d\tau/d\dot{\gamma} \tag{5}$$

The apparent viscosity is greater than the consistency.

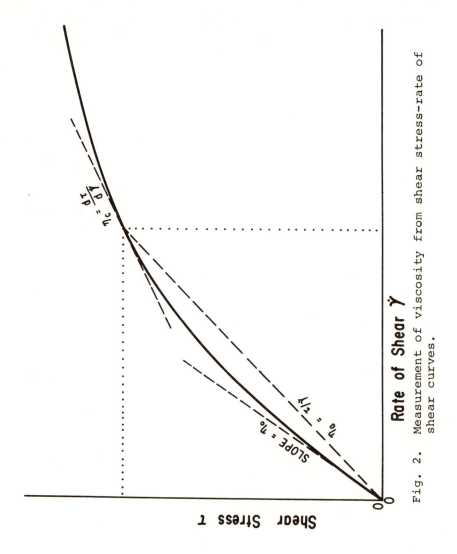

Fig. 2. Measurement of viscosity from shear stress–rate of shear curves.

In cases where the relative velocity of the shearing plates is not constant but varies in a sinusoidal
manner, a complex viscosity $\eta*$ is measured. The complex
viscosity contains an elastic component in addition to
a term similar to the ordinary steady state viscosity.
The complex viscosity is defined by:

$$\eta* = \eta' - i\eta'' \quad .$$

(6)

The dynamic viscosity η' is related to the steady state
viscosity and is the part of the complex viscosity that
measures the rate of energy dissipation. The imaginary
viscosity η'' measures the elasticity or stored energy
and is related to the shear modulus G' by

$$G' = \omega\eta''$$

(7)

where ω is the frequency of the oscillations in radians
per second, and $i = \sqrt{-1}$. The cause and manifestations
of elasticity in polymer melts will be discussed in
more detail in later chapters.

Another viscosity can be measured in tension instead of by shearing tests. The measurement of tensile
or extensional viscosity is illustrated in the lower
half of Figure 1 in which a strip or thread of material
is stretched. The tensile stress σ is

$$\sigma = \frac{F}{A}$$

(8)

while the tensile strain or elongation ε is

$$\varepsilon = \ln(L/L_0) \doteq \frac{L - L_0}{L_0}$$

(9)

L_0 is the initial length while L is the length at some

later time. The tensile viscosity η_t is

$$\eta_t = \frac{\sigma}{d\epsilon/dt} \,.$$ (10)

For Newtonian liquids, the tensile viscosity is three
times the shear viscosity, but for polymeric liquids
the tensile viscosity may be many times the shear
viscosity. For polymers, the tensile viscosity is
comparable in importance to the shear viscosity. Tensile viscosity is of great practical importance when
polymers flow through channels or tubes in which the
cross sectional area is decreasing. Examples include
the spinning of fibers and the filling of molds in
injection molding.

IV. NORMAL STRESSES

Normal stresses are other rheological phenomenon
encountered with non-Newtonian fluids. These normal
stresses generally develop at right angles to F when
the material is sheared. The first normal stress
difference $(\sigma_{11} - \sigma_{22})$ tends to force the shear plates
apart. The second and third normal stress differences
tend to create bulges in the polymer at the edge of the
plates either parallel or perpendicular to the direction
of the applied force F. The normal stresses are schematically shown in Figure 3 for a volume element inside
the fluid being sheared. Experimentally, differences
between two normal stresses are measured. The following definitions and relationships apply to normal
stresses:

$$\sigma_{11} + \sigma_{22} + \sigma_{33} = 0$$ (11)

$$\sigma_{11} - \sigma_{22} = \text{First normal stress difference}$$ (12)

$$\sigma_{22} - \sigma_{33} = \text{Second normal stress difference.}$$ (13)

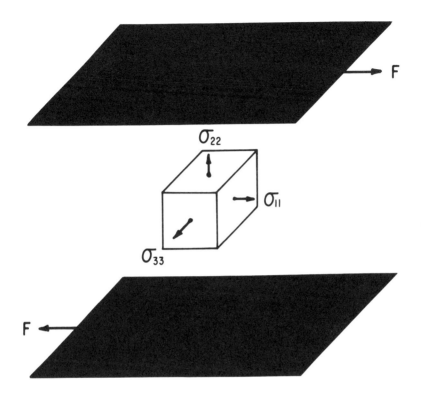

Fig. 3.   Schematic diagram showing the notation
          for normal stresses in a shear field.

Normal stresses and related phenomena are discussed in
a later chapter.

V.   GENERAL REFERENCES

1.   E. C. Bernhardt, Processing of Thermoplastic
     Materials, Reinhold, New York, 1959.

2.   J. A. Brydson, Flow Properties of Polymer Melts,
     Van Nostrand Reinhold, New York, 1970.

3.   F. R. Eirich, Rheology, Vol. 1-5, Academic Press
      New York.

4.   C. D. Han, Rheology in Polymer Processing, Academic
      Press, New York, 1976.

5.   R. S. Lenk, Plastics Rheology, Interscience,
      New York, 1968.

6.   J. M. McKelvey, Polymer Processing, John Wiley,
      New York, 1962.

7.   R. A. Mendelson, Melt Viscosity, Encyclopedia of
      Polymer Science and Technology, Vol. 8, p. 587,
      Interscience, New York, 1968.

8.   S. Middleman, The Flow of High Polymers,
      Interscience, New York, 1968.

9.   J. R. A. Pearson, Mechanical Principles of Polymer
      Melt Processing, Pergamon Press, Oxford, 1966.

10.   E. T. Severs, Rheology of Polymers, Reinhold,
      New York, 1962.

11.   J. R. Van Wazer, J.W. Lyons, K. Y. Kim, and
      R. E. Colwell, Viscosity and Flow Measurement,
      Interscience, New York, 1963.

Chapter 2

INSTRUMENTS

I.   INTRODUCTION   11
II.   CAPILLARY VISCOMETERS   12
III. COAXIAL CYLINDER VISCOMETERS   18
IV.   CONE AND PLATE RHEOMETERS   20
V.   PARALLEL PLATE VISCOMETERS   22
VI.   TENSILE OR EXTENSIONAL VISCOMETRY   23
VII. DYNAMIC OR OSCILLATORY RHEOMETERS   24
VIII. REFERENCES   27

I. INTRODUCTION

A great variety of instruments has been used to
measure the viscosity and other rheological properties
of liquid and molten polymers [1,2]. Most of these
instruments are capable of measuring the rheological
properties as a function of temperature and rate of
shear. These instruments can be put into a number of
classes. One of these classes is the steady state
instruments; most of the instruments used in the past
fall into this class. These instruments include the
simple shear viscometers, coaxial cylinder viscometers,
capillary rheometers, cone and plate viscometers, and
parallel plate viscometers.

Another class of instrument measures the complex
viscosity. Dynamic rheometers, rheogoniometers, and
orthogonal rheometers are in this class. In recent
years, this type of measurement has become very popular.
Another, less conventional, class of instruments measures tensile viscosity. Normal stress measurements
require yet another class of instruments. Some

instruments measure both viscosity and normal stresses.

II. CAPILLARY VISCOMETERS

A very popular type of instrument for studying the rheological behavior of molten polymers is the capillary viscometer [1-4]. As shown in Figure 1, a liquid polymer is forced by a piston or by pressure from a reservoir through a capillary. The quantity of polymer coming from the capillary per unit of time at a given pressure drop is the basic measurement used to calculate the viscosity.

The capillary rheometer has a number of advantages. First, the instrument is relatively easy to fill. This is an important consideration since most polymer melts are too viscous to pour readily even at high temperatures. The test temperature and rate of shear are varied readily. The shear rates and flow geometry are similar to the conditions actually found in extrusion and in injection molding. In addition to the viscosity, some indication of polymer elasticity is found from the die swell of the extrudate. Finally, factors affecting the surface texture of the extrudate and the phenomenon of melt fracture can be studied.

The main disadvantage of the capillary viscometer is that the rate of shear is not constant but varies across the capillary. Another disadvantage is the necessity of making a number of corrections in order to get accurate viscosity values.

The important equations pertaining to the capillary viscometer for Newtonian fluids are:

$$\eta = \frac{\pi R^4 P}{8LQ} \tag{1}$$

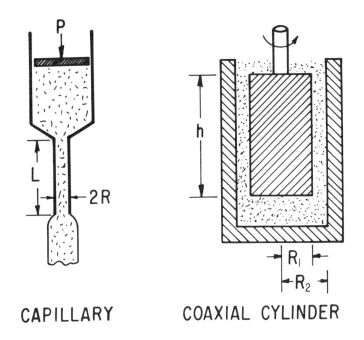

CAPILLARY   COAXIAL CYLINDER

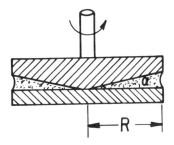

CONE AND PLATE

Fig. 1.   Schematic diagrams of three types of
          viscometers.

$$Q = V/t \; ; \; \tau_w = \frac{RP}{2L} \tag{2}$$

$$\dot{\gamma}_w = \frac{4Q}{\pi R^3} = \frac{RP}{2L\eta} = \frac{4\bar{v}}{R} \; ; \quad \bar{\dot{\gamma}} = \frac{8Q}{3\pi R^3} \tag{3}$$

$$\bar{v} = \frac{R^2 P}{8L\eta} \; ; \qquad \eta = \frac{\tau_w}{\dot{\gamma}_w} \tag{4}$$

In these equations, η is the viscosity, R is the radius of the capillary which has a length L, Q is the volumetric flow rate through the capillary under a pressure drop P along the capillary, V is the total volume of fluid extruded in a time interval t, and $\dot{\gamma}$ is the shear rate $d\gamma/dt$. The rate of shear at the wall of the capillary is $\dot{\gamma}_w$ while $\bar{\dot{\gamma}}$ is the average rate of shear across the capillary. The shear stress at the wall is τ_w, and $\bar{v}$ is the average velocity across the capillary cross sectional area. The shear rate is a maximum at the wall, but the velocity is zero at the wall. The characteristics of capillary flow are shown in the bottom section of Figure 2. The solid lines are for Newtonian fluids. The dashed lines are typical of polymeric non-Newtonian liquids. Under conditions in which Q is the same for both Newtonian and non-Newtonian fluids, the area under the velocity distribution curves should be the same rather than as shown in Figure 2 where conditions have been adjusted to match the velocities at the center of the capillary.

Two corrections are commonly applied to capillary data in order to obtain the correct viscosity of polymeric fluids. The Rabinowitsch equation corrects the rate of shear at the wall for non-Newtonian liquids [5,6].

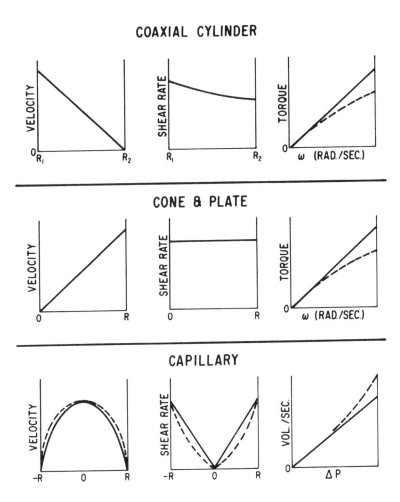

Fig. 2.   Characteristics of three types of viscometers.
Solid lines refer to Newtonian fluids while
dotted lines are typical of non-Newtonian
polymer melts.

It changes Equation 3 to

$$\dot{\gamma}_w = \frac{4Q}{\pi R^3} \left(\frac{3n + 1}{4n} \right) , \tag{5}$$

$$n = \frac{d \log \tau}{d \log \dot{\gamma}} . \tag{6}$$

The factor n equals 1.0 for Newtonian liquids, and it is
a constant for non-Newtonian fluids if the liquid obeys
the power law in which the shear stress depends upon $\dot{\gamma}$
to the nth power. Otherwise, n is a function of the
shear rate.

The Bagley [7,8] correction takes care of non-
ideality arising from viscous and elastic effects at the
entrance to the capillary. The effective length of a
capillary is greater than its true length. The shear
stress at the wall of equation 2 becomes

$$\tau_w = \frac{RP}{2(L + eR)} = \frac{P}{2(L/R + e)} = \frac{P-P_o}{2L/R} \tag{7}$$

The Bagley correction factor e should be independent of
capillary length, but in general it does vary somewhat
with L/R because of elasticity of polymer melts. The
Bagley correction is determined by measuring the pressure drop ΔP at constant rate of shear for several
capillary lengths and extrapolating to zero pressure
drop as shown in Figure 3. In Equation 7, P_o is the
pressure drop corresponding to a capillary of zero
length for a given rate of shear.

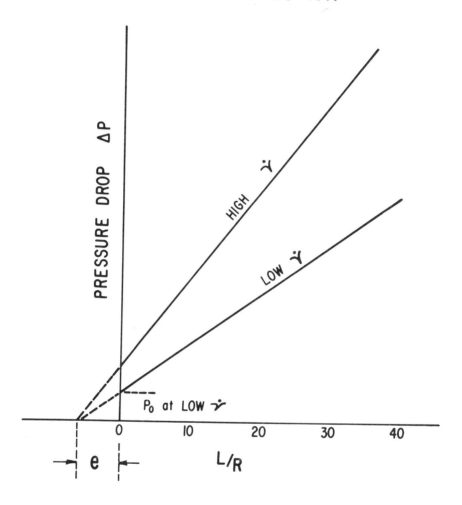

Fig. 3.   The Bagley correction for capillary rheometers.

III. COAXIAL CYLINDER VISCOMETERS

The coaxial cylinder or concentric cylinder viscometer is recognized as a basic instrument for measuring viscosity [1,2,9]. This instrument is used only
occasionally for high viscosity melts [10]. However, it
is a good instrument for low viscosity liquids, polymer
solutions, plastisols, and latices. In this instrument,
the fluid is placed in the annular space between two
cylinders. In one version of the apparatus, the inner
cylinder is rotated at a constant speed, while the
torque acting on either the outer or inner cylinder is
measured by a transducer. In some cases the instrument
is modified so that the outer cylinder is rotated while
the torque is measured by a transducer on the shaft of
the inner cylinder. In another version of the viscometer, a constant torque is applied to the inner
cylinder while its speed of rotation is measured by some
kind of a tachometer device.

A major advantage of the coaxial cylinder viscometer is the nearly constant shear rate throughout the
entire volume of fluid being measured if the space between cylinders is small. This is an important factor
with non-Newtonian polymer melts in which the viscosity
may be strongly dependent upon the rate of shear.
Coaxial cylinder instruments are calibrated easily, and
the corrections can be small. A major disadvantage of
these instruments is the difficulty in filling them
with a very viscous polymer melt. A less important, but
annoying, disadvantage is the creeping of the polymer
up the shaft of the inner cylinder because of normal
stresses developed in the polymer by the rotation of the
cylinder. Most coaxial cylinder instruments are limited
to fairly low rates of shear which may be less than the

rates of shear encountered in practical processing
situations.

For Newtonian liquids, the important equations for
coaxial cylinder viscometers (neglecting end corrections)
are:

$$\eta = \frac{M}{4\pi h\omega} \left(\frac{1}{R_1^2} - \frac{1}{R_2^2} \right) = \frac{\tau_x}{\dot{\gamma}_x} \tag{8}$$

$$\tau_x = \frac{M}{2\pi R_x^2 h} \quad ; \quad R_1 \leq R_x \leq R_2 \tag{9}$$

$$\dot{\gamma}_x = \frac{2\omega R_1^2 R_2^2}{R_x^2 \left(R_2^2 - R_1^2 \right)} \tag{10}$$

A torque M is produced by the rate of angular rotation ω
in radians per second. As shown in Figure 1, the radius
of the inner and outer cylinders is R_1 and R_2, respectively. The immersed length of cylinder is h while R_x
is any arbitrary radius between R_1 and R_2. The top
part of Figure 2 illustrates the characteristics of
coaxial cylinder instruments. The solid curves are for
Newtonian liquids while the dashed line is typical of a
non-Newtonian liquid. If the annular spacing is small
compared to the diameter of the cylinders, the shear
rate is nearly constant across the annular gap.

The main correction that must be applied to data in
order to get the correct viscosity is an end correction
for a coaxial cylinder viscometer. The liquid at the
ends of the inner cylinder create additional drag and
give rise to a torque in addition to the torque due to
the liquid in the annular gap. Thus, the measured

torque is equivalent to that of an inner cylinder with
an apparent length greater than its true length. Equation 8 can be replaced by

$$\eta = \frac{M}{4\pi\omega(h + h_o)} \left(\frac{1}{R_1^2} - \frac{1}{R_2^2} \right) \tag{11}$$

where h_o is a correction length. The correction h_o can
be evaluated by changing the immersion depth and extrapolating to a zero depth of immersion. However, in most
cases it is easier to calibrate a viscometer with a
liquid of known viscosity and to use the following
equation:

$$\eta = \frac{kM}{\omega} . \tag{12}$$

The instrument constant k takes into account any corrections as long as the volume of liquid in the viscometer
is held constant.

IV. CONE AND PLATE RHEOMETERS

The geometrical characteristics of a cone and plate
rheometer are shown in the bottom section of Figure 1.
A flat circular plate and a linearly concentric cone are
rotated relative to each other. The liquid is in the
space between the plate and cone. This type of rheometer has become very popular with rheologists who work
with viscous polymer melts [11-13]. The general characteristics of cone and plate rheometers are shown in
Figure 2.

A major advantage of a cone and plate rheometer is
the constant shear rate throughout all the liquid. As
pointed out before, this constancy is especially important with molten polymers. This instrument also has

other advantages. The sample size is very small, so the
instrument is valuable for evaluating the rheological
behavior of experimental samples where only small quantities of material are available. The small sample size
also results in less heat build-up at high rates of shear
than with a coaxial cylinder apparatus. Another advantage is the ease of loading the sample and the ease of
cleaning the apparatus at the end of a test. The instrument can be modified to measure normal stresses in
addition to the viscosity. The instrument has a number
of disadvantages: The rate of shear is limited to rather
low rates of shear. There is a tendency for the development of secondary flows in the polymer, and for the
polymer to crawl out of the instrument or to break up in
such a manner that accurate measurements become impossible at high rates of shear. Accurate spacing of the
cone and plate is required, so that more experimental
skill may be required than for some other instruments.
    For Newtonian liquids, the basic equations for cone
and plate rheometers are:

$$\eta = \frac{3\alpha M}{2\pi R^3 \omega} \qquad (13)$$

$$\dot{\gamma} = \omega/\alpha \qquad (14)$$

$$\tau = \frac{3M}{2\pi R^3} \qquad (15)$$

As shown in Figure 1, R is the radial distance of the
sample in the cone and plate, and α is the angle in
radians which the cone makes with the flat plate. M is

the torque required to rotate the cone relative to the
plate at an angular frequency of ω radians per second.

The cone and plate rheometer can be used to measure
directly the normal stress which tends to force the cone
and plate apart during rotation. The difference between
the first and second normal stresses is [14]

$$\sigma_{11} - \sigma_{22} = \frac{2N}{\pi R^2} \tag{16}$$

where N is the normal (axial) force trying to separate
the cone and plate. The quantity $(\sigma_{11} - \sigma_{22})$ is called
the first normal stress difference. Normal stresses will
be discussed in more detail in a later chapter.

V. PARALLEL PLATE VISCOMETERS

In the parallel plate viscometer, a circular disk
of viscous liquid is compressed between two parallel
plates which are larger in diameter than the diameter
of the liquid disk [1,2,15]. The force F is applied
perpendicular to the plates, and the spacing of the
plates is measured as a function of time t. For
Newtonian liquids when there is no slippage of the fluid
at the faces of the plates, the viscosity η is given by

$$\frac{1}{\eta} = \frac{3V^2}{8\pi Ft} \left(\frac{1}{h^4} - \frac{1}{h_o^4} \right). \tag{17}$$

The volume of the fluid is V, h_o is the initial spacing
of the plates, and h is the spacing at time t. The
parallel plate viscometer is suited best for liquids
with extremely high viscosities deformed at low rates of
shear. The instrument has the disadvantage that the
shear rate is generally unknown.

VI. TENSILE OR EXTENSIONAL VISCOMETRY

With the growing realization of the importance of extensior 1 or tensile viscosity in many commercial fabrication processes, such measurements have recently become quite popular. A number of different types of instruments have been used. As suggested in the bottom half of Figure 1 of Chapter 1, some instruments are very similar to those used for measuring the stress-strain properties of rigid polymers [16-20]. The two ends of a bar of molten polymer are attached to cooled clamps and separated at a rate so as to maintain a constant rate of tensile strain, $\dot{\varepsilon} = d \ln(L/L_0)/dt$. The extensional viscosity η_t is calculated from the force F required to stretch the polymer by

$$\eta_t = \frac{F/A}{\dot{\varepsilon}} = \frac{\sigma}{\dot{\varepsilon}} \ . \qquad (18)$$

Similar types of instruments stretch the polymer strand issuing from a capillary rheometer.

Meissner [21] stretches a strand of molten polymer by placing it between two sets of rollers, similar to a mill roll, which rotate in opposite directions. Other instruments create the extensional flow by squirting the polymer from a large reservoir through an orifice or by flowing the polymer through a converging duct [18,22].

The above instruments create an uniaxial extensional flow. In the blowing of films and in some similar processes, biaxial extensional flow is important. Denson and Gallo [23] describe an instrument for measuring this biaxial tensile viscosity.

VII. DYNAMIC OR OSCILLATORY RHEOMETERS

Ferry [24] describes many of the instruments used
for measuring the complex viscosity. The significance of
such data and the interrelations between complex viscosity terms is discussed by Ferry [24] and Nielsen [25].
The basic essentials of most such instruments are schematically illustrated in Figure 4. The polymer sample is
deformed in shear or in tension by some oscillating driv-
er, which may be either mechanical or electromagnetic in
nature. The amplitude of the sinusoidal deformation is
measured by a strain transducer which may be a linear
variable differential transformer (LVDT), a variable
resistance gage, or some type of optical transducer. The
force deforming the sample is measured by the small deformation of a relatively rigid spring or torsion bar to
which is attached a stress transducer. The stress transducer most often is a resistance strain gage or another
LVDT. Because of energy dissipated by the viscous fluid,
a phase difference develops between the stress and the
strain. The complex viscosity behavior is determined
from the amplitudes of the stress and strain and the
phase angle between them. A few of the commercial instruments and others which are widely used are described
in the following references [11,13,26-32]. A special
type of oscillating rheometer, known as the orthogonal
rheometer, consists of two parallel disks the space between which is filled with polymer and whose axes are not
quite colinear. As one disk rotates relative to the
other, an eccentric oscillatory motion is set up in the
polymer [33].

Dynamic rheometers have a great advantage over most
other types of rheometers because such instruments
measure the elastic modulus of polymer melts in addition
to the viscosity. In many practical fabrication applica-

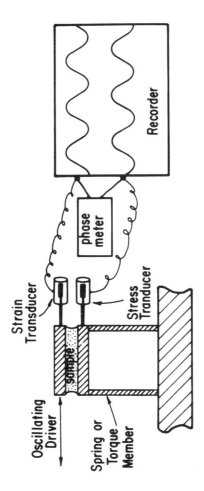

Fig. 4.   Schematic diagram of an oscillating
rheometer for measuring the dynamic
properties of fluids.

tions the elasticity of melts is comparable in importance
to the viscosity. The non-Newtonian behavior of polymer
melts is largely due to their elasticity. Most dynamic
rheometers are capable of operating over a wide frequency
range. Angular frequency for a dynamic rheometer is
directly analogous to rate of shear for a steady state
viscometer. A possible disadvantage of most dynamic
rheometers is the small amplitude of deformation. In
most fabrication processes, the polymer undergoes very
large deformations.

In dynamic or oscillatory rheometers the strain and
stress in shear are:

$$\gamma = \gamma_o \sin \omega t \qquad (19)$$

$$\tau = \tau_o \sin (\omega t + \delta) \qquad (20)$$

$$|\eta^*| = \sqrt{\eta'^2 + \eta''^2} \ = \frac{\tau_o}{\dot{\gamma}_o} \qquad (21)$$

$$\eta^* = \eta' - i\eta'' \qquad (22)$$

The maximum values of the sinusoidal shear strain and
shear stress are γ_o and τ_o respectively. The complex
viscosity is η^*, η' is the dynamic viscosity or the real
part of the viscosity, and η'' is the imaginary part of
the viscosity, while $i = \sqrt{-1}$. Analogous equations
apply to extensional viscosity. The angular frequency
is ω, t is time, and δ is the phase angle between the
stress and strain, that is,

$$\tan \delta = \frac{\eta'}{\eta''} \qquad (23)$$

if the frequency of the oscillating system is not near a
resonance frequency. The usual steady state shear viscosity is closely related to η' while η'' is an elasticity
term related to the dynamic shear modulus G' of the
polymer by

$$\eta'' = G'/\omega \ . \qquad (24)$$

Other relations which are sometimes useful are:

$$G^* = i\omega\eta^* \qquad (25)$$

$$G' = \omega\eta'' \qquad (26)$$

$$\tan \delta = G''/G' \qquad (27)$$

$$\eta' = |\eta^*| \sin \delta \; ; \; \eta'' = |\eta^*| \cos \delta. \qquad (28)$$

In later chapters, the relationships between dynamic viscosity and steady state viscosity will be discussed.

VIII. REFERENCES

1. J. R. Van Wazer, J. W. Lyons, K. Y. Kim, and R. E. Colwell, Viscosity and Flow Measurement, Interscience, New York, 1963.

2. S. Oka, Rheology Vol. 3, p. 17, F. R. Eirich, Ed., Academic Press, New York, 1960.

3. H. K. Nason, J. Appl. Phys., 16, 338 (1945).

4. E. H. Merz and R. E. Colwell, ASTM Bull., 232, 63 (1958).

5. B. Rabinowitsch, Z. Physik. Chem. (Leipzig), 145A, 1 (1929).

6. I. M. Krieger and S. H. Maron, J. Appl. Phys., 23, 147 (1952).

7. E. B. Bagley, J. Appl. Phys., 28, 624 (1957).

8. E. B. Bagley, Trans. Soc. Rheol., 5, 355 (1961).

9. R. Buchdahl and J. E. Thimm, J. Appl. Phys., 16, 344 (1945).

10. R. Buchdahl, J. Colloid Sci., 3, 87 (1948).

11. K. Weissenberg, Proc. 1st Internat. Congr. Rheol., p. 114, North Holland Publ., 1949.

12. W. F. O. Pollett and A. H. Cross, Rev. Sci. Instr.,
    27, 209 (1950).

13. C. Macosko and J. M. Starita, SPE J., 27, #11,
    38 (1971).

14. J. F. Petersen, R. Rautenbach, and P. Schümmer,
    Rheol. Acta, 14, 968 (1975).

15. G. J. Dienes and H. F. Klemm, J. Appl. Phys., 17,
    458 (1946).

16. J. M. Dealy, Polymer Eng. Sci., 11, 433 (1971).

17. R. L. Ballman, Rheol. Acta, 4, 137 (1965).

18. F. N. Cogswell, Rheol. Acta, 8, 187 (1969).

19. C. D. Denson, Polymer Eng. Sci., 13, 125 (1973).

20. G. V. Vinogradov, B. V. Radushkevich, and
    V. D. Fikhman, J. Polymer Sci., A2, 8, 1 (1970).

21. J. Meissner, Rheol. Acta, 8, 78 (1969).

22. A. B. Metzner and A. P. Metzner, Rheol. Acta, 9,
    174 (1970).

23. C. D. Denson and R. J. Gallo, Polymer Eng. Sci.,
    11, 174 (1971).

24. J. D. Ferry, Viscoelastic Properties of Polymers,
    John Wiley, New York, 2nd Ed., 1970.

25. L. E. Nielsen, Mechanical Properties of Polymers
    and Composites, Vol. 1, Marcel Dekker, New York,
    1974.

26. H. Markovitz, P. Yavorsky, R. Harper, L. Zapas, and
    T. DeWitt, Rev. Sci. Instr., 23, 430 (1952).

27. D. O. Miles, J. Appl. Phys., 33, 1422 (1962).

28. M. H. Birnboim and J. D. Ferry, J. Appl. Phys., 32,
    2305 (1961).

29. D. J. Plazek, M. N. Vrancken, and J. W. Berge,
    Trans. Soc. Rheol., 2, 39 (1958).

30. R. S. Marvin, E. R. Fitzgerald, and J. D. Ferry,
    J. Appl. Phys., 21, 197 (1950).

31. M. Yoshino and M. Takayanagi, J. Japan Soc. Test.
     Mater., 8, 330 (April 1959).

32. M. H. Birnboim and L. J. Elyash, Bull. Amer. Phys.
     Soc., Ser. II, 11, 165 (1966).

33. B. Maxwell and R. P. Chartoff, Trans. Soc. Rheol.,
     9, 41 (1965).

Chapter 3
EFFECTS OF TEMPERATURE AND PRESSURE

I.   TEMPERATURE DEPENDENCE OF VISCOSITY   31
II. MASTER CURVES FOR TEMPERATURE DEPENDENCE   36
III. EFFECT OF PRESSURE ON VISCOSITY   42
IV. REFERENCES   44

I. TEMPERATURE DEPENDENCE OF VISCOSITY

The viscosity of most polymers changes greatly with temperature. For Newtonian liquids and for polymer fluids at temperatures far above the glass transition temperature T_g or the melting point, the viscosity follows the Andrade or Arrhenius equation to a good approximation:

$$\eta \doteq K\ e^{E/RT} \ . \tag{1}$$

In this equation, K at a given shear stress is a constant characteristic of the polymer and its molecular weight, E is the activation energy for the flow process, R is the gas constant, and T is the temperature in degrees Kelvin. The activation energy generally is in the range between 5,000 and 50,000 cal/mole (2.09×10^7 to 2.09×10^8 joules/kg mole). Table 1 lists some typical values of the energy of activation for several polymers. The value of 4 kcal/mole for dimethyl silicone polymers is the smallest known; this low value is a result of the great flexibility of the silicone polymer chain. The energy of activation for flow increases as the size of side groups increases and as the chain becomes more rigid [1,2]. A simple example will illustrate the temperature dependence of viscosity. If E = 20,000

31

TABLE 1

Energy of Activation for Flow of Polymers

Polymer	Energy of Activation E	
	kcal/g mol	kJ/g mol
Dimethyl silicone	4	16.7
Polyethylene (High Density)	6.3-7.0	26.3-29.2
Polyethylene (Low Density)	11.7	48.8
Polypropylene	9.0-10.0	37.5-41.7
Polybutadiene (cis)	4.7-8	19.6-33.3
Polyisobutylene	12.0-15.0	50-62.5
Polyethylene terephthalate	19	79.2
Polystyrene	25	104.2
Poly(α methyl styrene)	32	133.3
Polycarbonate	26-30	108.3-125
Poly(1-butene)	11.9	49.6
Polyvinyl butyral	26	108.3
SAN (styrene acrylonitrile copolymer)	25-30	104.2-125
ABS (20% rubber) (acrylonitrile-butadiene-styrene copolymer)	26	108.3
ABS (30% rubber)	24	100
ABS (40% rubber)	21	87.5

cal/mole, changing the temperature from 300°K to 310°K
will decrease the viscosity by a factor of about 2.96.

The value of the energy of activation depends
strongly on whether the viscosities at various

temperatures are evaluated at constant shear stress or
at constant shear rate. If E is evaluated at a constant
shear stress, it is found that E is essentially a
constant independent of what value is chosen for τ.
However, if E is evaluated at a constant shear rate, the
energy of activation generally decreases with increasing
rate of shear [3].

For amorphous polymers at temperatures less than
100°C above their T_g, the Andrade equation does not fit
the data well. A much better equation is the Williams-
Landel-Ferry or the W-L-F equation [4,5]:

$$\log_{10}\frac{\eta}{\eta_g} \doteq -\left[\frac{17.44\left(T - T_g\right)}{51.6 + T - T_g}\right] - \log_{10}\left(\frac{T_g\ \rho_g}{T\ \rho}\right) \quad (2)$$

$$T_g \leq T \leq T_g + 100$$

The viscosity at T_g at low rates of shear is η_g, T is
the temperature in degrees Kelvin, ρ is the density at
that temperature, and ρ_g is the density at T_g. The
term $\log\left(\frac{T_g\ \rho_g}{T\ \rho}\right)$ is generally near zero. The viscosity
at T_g is often about 10^{13} poises. Figure 1 illustrates
how the viscosity changes with temperature according
to the W-L-F equation. The Andrade equation would give
nearly a straight line on the same type of plot. If the
viscosity and its temperature coefficient are the same
for the W-L-F and Andrade equations at 100°C above T_g,
the viscosity according to the Andrade equation would be
given approximately by the dashed line in Figure 1. The
Andrade equation is generally satisfactory if the temperature is greater than $(T_g + 100°)$; at lower temperatures, the W-L-F or similar type of equation should be
used with an adjustable empirical value for η_g.

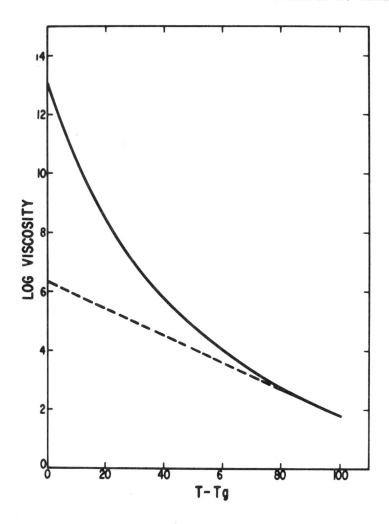

Figure 1. Viscosity as a function of temperature
according to the W-L-F equation. Dotted
line gives the approximate viscosity
dependence according to the Arrhenius
equation when matched to the W-L-F value
at $(T_g + 100°)$.

For liquids obeying the Andrade equation, the
energy of activation E is nearly independent of the
temperature. However, for the W-L-F equation, the

apparent energy of activation or the temperature coefficient of viscosity not only depends upon temperature
but also upon the glass transition temperature T_g. For
the W-L-F equation at low rates of shear,

$$E = \frac{Rd(\log \eta)}{d(1/T)} = \frac{4.12 \times 10^3 T^2}{(51.6 + T - T_g)^2} . \tag{3}$$

Table 2 gives E for various values of T_g and $(T - T_g)$.

TABLE 2
W-L-F Energy of Activation for Viscous Flow

$T-T_g$	Energy of Activation (kcal/mole)			
	T_g=200°K	T_g=250°K	T_g=300°K	T_g=350°K
0	61.9	96.7	139.3	189.6
2	58.5	91.1	130.8	177.7
5	54.1	83.6	119.6	162.1
10	47.9	73.4	104.3	140.7
20	38.9	58.6	82.3	110.0
30	32.7	48.5	67.4	89.4
50	25.0	35.9	48.9	63.9
80	18.7	25.9	34.4	44.0
100	16.1	22.0	28.7	36.3

The energy of activation becomes very large as the temperature approaches T_g, especially if T_g is large.
    The energy of activation can be estimated by as-
signing values to each of the groups making up the monomeric unit and adding up the values for all the
groups [6]. The energy of activation E is given by

$$E = \left(\frac{\Sigma X_i}{M_o}\right)^3 . \tag{4}$$

The molecular weight of the monomeric unit is M_O, and X_i is an empirical value assigned to group i. Values of X_i have been tabulated by van Krevelen and Hoftyzer [6].

II. MASTER CURVES FOR TEMPERATURE DEPENDENCE

Because of the great sensitivity of polymer viscosity to both temperature and shear rate, a large amount of data is required to characterize the flow behavior of a polymer. Thus, there is a great need to predict the viscosity from a small amount of experimental data. There also is the need to be able to compare the behavior of one polymer with other polymers. Several schemes have been developed for doing this by superposition rules in which one curve is shifted relative to another curve. The superimposed curves form a single curve, which is called the master curve [5].

Mendelson [7-9] has proposed a method for the prediction of flow behavior at various temperatures by means of a shear rate-temperature superposition method. Curves of shear stress τ are plotted as a function of shear rate $\dot{\gamma}$ as in Figure 2 for several temperatures. (Note that logarithmic scales are used.) The apparent viscosity can be calculated from such curves at a given shear rate by

$$\eta = \tau_w / \dot{\gamma}_w \tag{5}$$

The curve at one convenient temperature can be chosen as the reference curve. In Figure 2 the curve at 200°C was selected as the reference curve. All the other curves can then be superimposed on the reference curve by shifting them horizontally along the shear rate axis. Curves for temperatures above the reference temperature

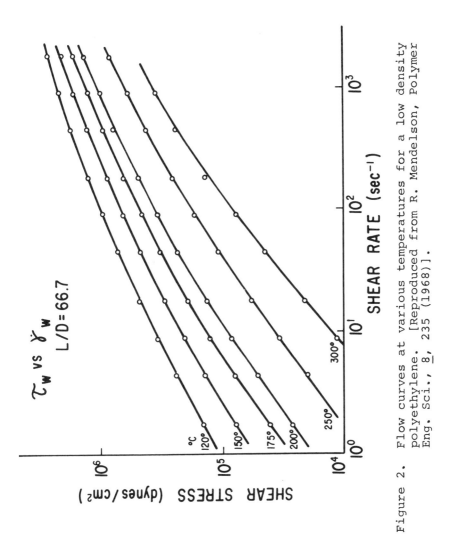

Figure 2.   Flow curves at various temperatures for a low density polyethylene. [Reproduced from R. Mendelson, Polymer Eng. Sci., 8, 235 (1968)].

are shifted to the left while curves for lower temperatures are shifted to the right. The resulting master
curve is shown in Figure 3. It is important to know how
much each curve had to be shifted to superpose upon the
master curve and how this shifting depended upon the
temperature. The amount of shifting along the log $\dot{\gamma}$
axis is called the shift factor, a_T. A shift factor of
10 means that the curve is shifted one decade along the
log $\dot{\gamma}$ scale while if $a_T = 100$, the curve is shifted two
decades along the log $\dot{\gamma}$ scale. Mathematically, where

$$a_T = \frac{\dot{\gamma}(\text{reference})}{\dot{\gamma}(T)} \qquad (6)$$

$\dot{\gamma}$ (reference) is the shear rate for a constant value of
shear stress for the reference temperature curve, and
$\dot{\gamma}(T)$ is the shear rate at temperature T for the same
value of the shear stress. It has been shown that the
shift factor also may be given by

$$a_T = \frac{\eta(T)}{\eta(\text{reference})} = K_1 e^{E/RT} \qquad (7)$$

where K_1 is a constant, and E is the energy of activation for flow evaluated at constant shear stress. The
procedure can be reversed by using the master curve to
calculate the flow behavior at any arbitrary temperature
or to estimate the behavior of a polymer of different
molecular weight but similar structure with a minimum of
information. This is a very powerful technique since
for many polymers the shift factors are independent of
molecular weight [7]. From the constructed τ versus $\dot{\gamma}$
curve, the viscosity as a function of shear rate may be
calculated by Equation 5.

    An alternate technique for predicting the flow be-
havior at different temperatures is to first calculate

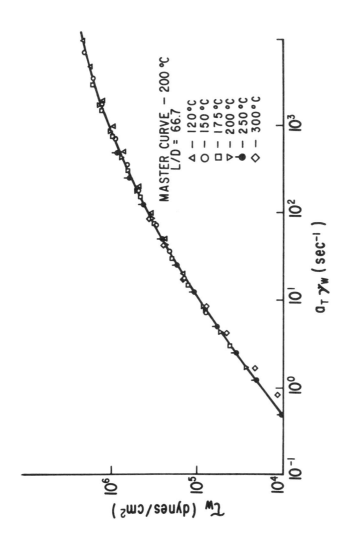

Figure 3.   Master curve produced from the data of Figure 2 for a
low density polyethylene. Reference temperature =
200°C. [Modified from R. Mendelson, Polymer Eng. Sci.,
8, 235 (1968)].

the viscosity at a given temperature as a function of
shear rate before shifting the curves to make a master
curve [10-16]. A set of such curves is shown in Figure 4.
Most polymer melts become Newtonian at very low rates of
shear and have a zero shear viscosity, η_o, which is a
function of temperature. The viscosity-shear rate curves
are first normalized by dividing the viscosity by η_o to
give a relative viscosity η/η_o. The relative viscosity
versus shear rate curves, which were obtained at various

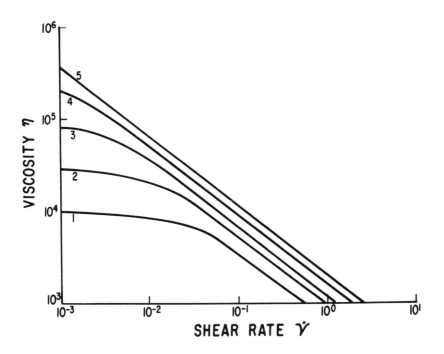

Figure 4. Typical log viscosity-log shear rate curves
           at five different temperatures. Curve 1 is
           for the highest temperature, and curve 5 is
           for the lowest temperature. For a typical
           polymer, the temperature difference between
           each curve is approximately 10°C.

temperatures, then can be superimposed to a single master curve if log η/η_o is plotted agains log $(\dot{\gamma}\eta_o)$. A typical example is shown in Figure 5, which was constructed from Figure 4. In some cases, better fit of the data is obtained if the abscissa scale is log $(\dot{\gamma}\eta_o/T)$ instead of log $(\dot{\gamma}\eta_o)$; note that $\dot{\gamma}\eta_o = \tau$ in the Newtonian region. Therefore, the master curve can be made by plotting log η/η_o versus log τ. The effect of temperature largely is automatically compensated for by using the relative viscosity η/η_o instead of the viscosity itself and by replacing the shear rate with the shear stress τ. In some cases this technique is difficult to use because η_o at a given temperature may be unknown. For example, in

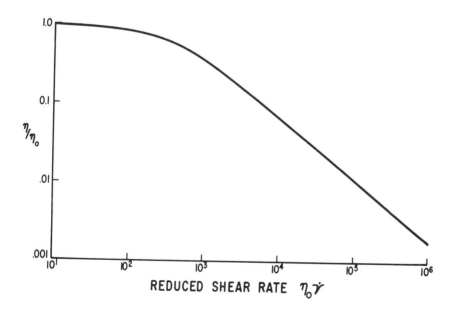

Figure 5.   The master curve produced from the data of Figure 4 by vertical and horizontal shifting of the curves.

Figure 4, η_o is obvious only for curves 1 and 2, while for curves 4 and 5 it becomes impossible to properly extrapolate to η_o. For these cases, η_o may be obtained from alternate data such as from the known temperature dependence of η_o.

III. EFFECT OF PRESSURE ON VISCOSITY

Several theories suggest that the viscosity of a fluid is determined by its free volume [17-21]. The free volume of a liquid is defined in various ways, but a common definition is the difference between the actual volume and a volume in which such close packing of the molecules occurs that no motion can take place. The greater the free volume the easier it is for flow to take place. Free volume increases with temperature because of thermal expansion. However, the most direct influence on free volume should be the pressure. An increase in hydrostatic pressure decreases free volume and increases the viscosity of a liquid. It has been suggested that the Andrade equation for the temperature dependence of viscosity should be modified to [17,18]:

$$\eta = K \exp \left(\frac{E}{RT} + \frac{CV_o}{V_f} \right) .$$ (8)

The close packed volume is V_o, and V_f is the free volume, which is defined as $V-V_o$ where V is the observed volume. The constant C generally is between 0.5 and 1.0. The close packed volume V_o is approximately the volume at the glass transition temperature. In this case,

$$V_f \doteq \alpha V_o (T-T_g)$$ (9)

where α is the coefficient of thermal expansion.

The pressure can be quite high in extruders, injection molding machines, and in capillary rheometers. One suspects that the viscosity in such equipment could be higher than what one would measure in a cone and plate rheometer, for example. This higher viscosity due to pressure may exist, but the increased viscosity often is unnoticed because the increase may be partly offset by viscous heating of the polymer in the equipment [22]. The shear heating of the polymer in the equipment raises the temperature above the indicated value so that the viscosity may be lowered by an amount comparable to the increase due to the pressure. The high shear rate dependence of viscosity also obscures the effect of pressure [23].

There are not many data about the effect of pressure on the viscosity of polymer melts, and generally there is poor agreement between data from different sources. Viscosity either increases linearly with pressure or at a rate somewhat faster than the pressure. Westover [24] has described a double piston apparatus for measuring viscosity as a function of pressure up to 25,000 psi. He found the viscosity of polystyrene to increase by a factor of over a hundred times and the viscosity of polyethylene to increase by a factor of five at a constant rate of shear as the pressure was increased. Maxwell and Jung [25] found that the viscosity of polystyrene increased by a factor of 135 times when the pressure was increased from zero to 18,000 psi at 385°F. The viscosity of polyethylene increased by 14 times in the same pressure range at 300°F. However, Kamal and Nyun [26] found that the viscosity of polystyrene increased by only about a factor of two in going to a pressure of 600 atmospheres (8811 psi) at 180°C. They used a capillary rheometer and attempted to correct

the viscosity for viscous heating. A few workers have
measured the dynamic viscosity and shear modulus as a
function of pressure [27]. The dynamic viscosities of
low density polyethylene at 130° to 150°C and silicone
rubber at 25° to 80°C increased by a factor of about
three in going to a pressure of 1000 kg/cm^2 (14220 psi).
The dynamic shear modulus increased in about the same
proportion. Both log η' and log G' versus log ω curves
could be superimposed at all pressures to give master
curves by equal horizontal and vertical shifts.

IV. REFERENCES

1. H. Schott, J. Appl. Polymer Sci., 6, #3, S29 (1962).

2. R. S. Porter and J. F. Johnson, J. Polymer Sci.,
    C15, 373 (1966).

3. A. B. Bestul and H. V. Belcher, J. Appl. Phys., 24,
    696 (1953).

4. M. L. Williams, R. F. Landel, and J. D. Ferry,
    J. Amer. Chem. Soc., 77, 3701 (1955).

5. J. D. Ferry, Viscoelastic Properties of Polymers,
    2nd Ed., Wiley, New York, 1970.

6. D. W. van Krevelen and P. J. Hoftyzer, Ang. Makromol.
    Chem., 52, 101 (1976).

7. R. A. Mendelson, Polymer Eng. Sci., 8, 235 (1968).

8. R. A. Mendelson, Polymer Eng. Sci., 9, 350 (1969).

9. R. A. Mendelson, Trans. Soc. Rheol., 9, 53 (1965).

10. R. L. Ballman and R. H. M. Simon, J. Polymer Sci.,
    A2, 3557 (1964).

11. F. Bueche and S. Harding, J. Polymer Sci., 32, 177
    (1958).

12. G. Kraus and J. T. Gruver, Trans. Soc. Rheol., 13,
    315 (1969).

13. H. D. Herrmann and W. Knappe, Rheol. Acta, 8, 384 (1969).

14. C. K. Shih, Trans. Soc. Rheol., 14, 83 (1970).

15. A. Casale, R. S. Porter, and J. F. Johnson, J. Macromol. Sci., C5, 387 (1971).

16. D. P. Wyman, L. J. Elyash, and W. J. Frazer, J. Polymer Sci., A3, 681 (1965).

17. T. A. Litovitz and P. B. Macedo, Physics of Non-Crystalline Solids, J. Prins, Ed., Interscience, 1965, p. 220.

18. P. B. Macedo and T. A. Litovitz, J. Chem. Phys., 42, 245 (1965).

19. J. H. Hildebrand and R. H. Lamoreaux, Proc. Nat. Acad. Sci., 69, 3428 (1972).

20. I. C. Sanchez, J. Appl. Phys., 45, 4204 (1974).

21. M. Cohen and D. Turnbull, J. Chem. Phys., 31, 1164 (1959).

22. F. N. Cogswell, Plastics and Polymers, 41, 39 (1971).

23. J. F. Carley, Modern Plastics, 39, #4, 123 (Dec. 1961).

24. R. E. Westover, Polymer Eng. Sci., 6, 83 (1966).

25. B. Maxwell and A. Jung, Modern Plastics, 35, #3, 174 (Nov. 1957).

26. M. R. Kamal and H. Nyun, Trans. Soc. Rheol., 17, 271 (1973).

27. S. Tokiura, S. Ogihara, T. Takaki, and H. Sasaki, Proc. 5th Inter. Congr. Rheol., Vol. 4, p. 275, Univ. Tokyo Press, Tokyo, 1970.

Chapter 4

EFFECTS OF RATE OF SHEAR ON POLYMER RHEOLOGY

I.   GENERAL NON-NEWTONIAN BEHAVIOR OF   47
     POLYMER MELTS
II. ROLE OF ENTANGLEMENTS   48
III. THEORIES DESCRIBING THE SHEAR RATE   51
     DEPENDENCE OF VISCOSITY
IV. DYNAMIC PROPERTIES   52
V.   RHEOLOGY OF EMULSIONS, POLYBLENDS, AND   54
     BLOCK POLYMERS
     A. Emulsions   56
     B. Polyblends, Block and Graft Polymers   61
VI. REFERENCES   65

I. GENERAL NON-NEWTONIAN BEHAVIOR OF POLYMER MELTS

An outstanding characteristic of polymer melts is
their non-Newtonian behavior whereby the apparent viscosity η_a decreases as the rate of shear increases.
This viscosity decrease extends over many decades of
change in shear rate $\dot{\gamma}$, and the viscosity at high rates
of shear may be several orders of magnitude smaller than
the viscosity at low rates of shear. This non-Newtonian
behavior is of tremendous practical importance in the
processing and fabrication of plastics and elastomers.
First, the decreased viscosity makes the molten polymer
easier to process or squirt through small channels as in
the filling of a mold. At the same time, the energy
required to operate a large injection molding machine or

47

extruder is reduced by this same phenomenon. The Newtonian viscosity η_0 obtained at very low rates of shear
is misleadingly high if applied to commercial operations.
Second, the decrease in viscosity is associated with the
development of elasticity in the melt. This elasticity
produces such phenomena as die swell or the "puff-up" of
extruded strands. Molecular orientation in molded objects also is closely related to polymer elasticity.

Typical log η_a versus log $\dot{\gamma}$ curves are shown in
Figure 4 of Chapter 3. The melt has a Newtonian viscosity which is high at very low rates of shear. However,
over much of the usual accessible shear rate range, the
viscosity decreases nearly linearly with $\dot{\gamma}$ on the log-
log plot. In this linear range, the so-called power law
equation holds in which

$$\tau = K\dot{\gamma}^n \tag{1}$$

or

$$\eta_a = K\dot{\gamma}^{n-1} \tag{2}$$

where K and n are constants [1,2]. For Newtonian liquids, n = 1 and K = η_a; the value of n is less than one
for non-Newtonian polymer melts. The power law generally
does not hold accurately for more than two decades
change in shear rate, but in the range where the equation is valid, the slope of the straight line relationship on a log-log plot is:

$$\frac{d \log \eta}{d \log \dot{\gamma}} = n-1. \tag{3}$$

II. ROLE OF ENTANGLEMENTS

The basic cause for the non-Newtonian behavior of
polymer melts is the orientation of molecular segments
by the flow field. This orientation in turn gives rise

to elasticity in the melt. The elasticity is measured
as an elastic modulus. The orientation reduces the
entropy of the system in a manner analogous to the re-
duction in entropy in the kinetic theory of rubber elasticity [3-6]. Molecular entanglements with an appreciable lifetime exist above a critical molecular weight
Me [7-12]. As illustrated in Figure 1, entanglements

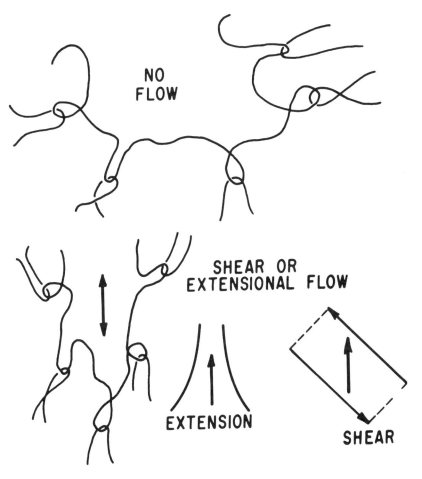

Figure 1.   Schematic diagram of how entanglements
            increase molecular orientation in flow
            fields.

greatly enhance the possibility of orienting molecular
segments in a flow field. The entanglements act as
temporary crosslinks, so that polymer melts may have
many of the characteristics of crosslinked rubbers. At
very low rates of shear, the entanglements have time to
slip and become disengaged before enough stress can develop in them to orient the molecules. At higher rates
of shear the segments between entanglements become
oriented before the entanglements can disappear. As a
load-bearing entanglement disappears, another entanglement which does not carry any load develops somewhere
else in the melt. Thus, there develops in the melt a
steady state condition in which the rates of formation
and destruction of entanglements become equal. As im-
plied in Figure 1, either shear flow or extensional flow
is capable of orienting polymer molecules. Extensional
flow is especially effective in creating elasticity in
melts in many cases. Elasticity and extensional flow
will be discussed in more detail in a later chapter.

From the above discussion, a polymer melt at rest
should have a higher concentration of entanglements than
a polymer which is flowing. The change in concentration
of entanglements has two effects: First, a polymer
should have a lower viscosity and less elasticity immedi-
ately after very high shear flow than it has after stand-
ing without flow for some extended time. This effect
has been observed experimentally [13-15]. Entanglements
not only increase elasticity, but they also increase the
viscosity because it becomes more difficult for flow to
occur by relative motion of the molecules when they are
entangled. This effect of entanglements is analogous to
the fact that it requires more force to pull a strand
from a ball of yarn when the strands are entangled and
knotted than when they are not entangled. Second, at

very high rates of shear practically no entanglements
can exist. At this point the elasticity should level
off. Also, the viscosity should reach a relatively
small value which becomes independent of the shear rate.
In other words, polymer melts may be expected to become
Newtonian in behavior at very high rates of shear. This
upper Newtonian region is normally not reached in practical situations because of viscous heating and flow instabilities.

III. THEORIES DESCRIBING THE SHEAR RATE DEPENDENCE OF VISCOSITY

There are several theories and resulting equations
which often are used to describe the shear rate dependence of the viscosity. These are the theories of
Bueche [16-18], Graessley [12,19,20], Williams [21,22],
and Cross [23]. Graessley and Bueche explicitly assume
there are molecular entanglements which decrease as the
rate of shear increases and which are forming and dis-
appearing in a dynamic steady state in a shear field.
Williams does not assume any given mechanism of intermolecular interaction in his theory, but he does assume
that the interactions are great enough to control the
shear stress. These theories are complex and do not
give simple equations for calculating the shear rate
dependence of the viscosity. On the other hand, the
Cross equation is a general empirical equation for fit-
ting curves which have a sigmoidal shape. The Cross
equation for the effect of shear rate on the apparent
viscosity η_a is:

$$\eta_a = \eta_\infty + \frac{\eta_o - \eta_\infty}{1 + \Omega\dot{\gamma}^m} . \qquad (4)$$

η_o and η_∞ are the limiting Newtonian viscosities at very

low and very high rates of shear, respectively. As
mentioned already, η_∞ may not be attainable experimentally, so it then must be considered to be an arbitrary
constant. Other constants are Ω and m. The Cross equation is shown graphically in Figure 2 for values of m
equal to 2/3 and 1.0 for the case where $\eta_0/\eta_\infty = 101$.
For monodispersed polymers, m $\doteq$ 1.0. For polymers with
a distribution of molecular weights, Cross [23] claims
that m $\doteq$ $(\overline{M}n/\overline{M}w)^{1/5}$. Therefore, for most polymers, m
should have values between 0.66 and 1.0. In the follow-
ing chapter, it will be shown that η_0 is determined by
the weight average molecular weight $\overline{M}w$ in addition to
the temperature for a given molecular structure. The
factors affecting η_∞ are not so well defined. However,
η_∞ does appear to increase with $\overline{M}w$ roughly to the first
power in some cases and to be nearly independent of
temperature [24]. In other cases, η_∞ is even less dependent on molecular weight.

The theories of Bueche, Graessley, and Williams
approximate the power law equation at the higher shear
rates. For monodisperse polymers the Bueche theory predicts that n = 1/2 in the power law equation, the
Graessley theory predicts that (n-1) equals - 9/11 while
the Williams theory predicts that n = 0. The published
experimental data show enough variation to include the
predictions of all the theories.

IV. DYNAMIC PROPERTIES

In dynamic or oscillatory measurements on polymer
melts, the angular frequency ω becomes analogous to the
rate of shear in the usual rheological measurements [25,
26]. As pointed out in Chapter 1, a dynamic shear modulus G' and a loss modulus or imaginary modulus G" are
measured. The loss modulus is related to the dynamic
viscosity η' by

$$G'' = \omega\eta' .$$

(5)

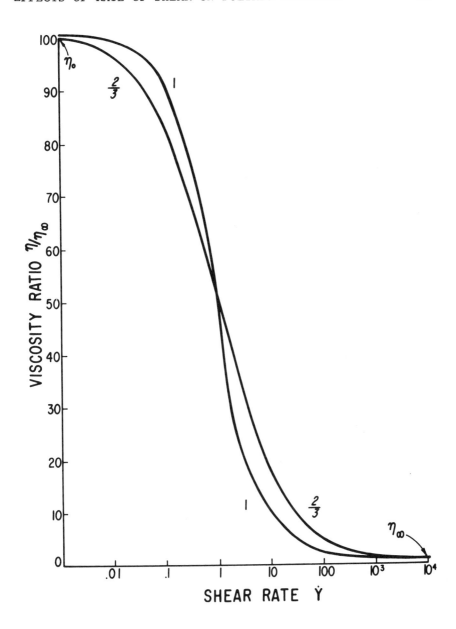

Figure 2. The apparent viscosity ratio η/η_∞ as a function of the shear rate according to the Cross equation for m = 2/3 and m = 1 when η_0/η_∞ = 101 and Ω = 1.

Figure 3 contains typical dynamic properties data
along with the more common steady-flow apparent viscosity as a function of shear rate [27]. In some cases G'
becomes greater than G" at high frequencies, and G" may
go through a broad maximum [28]. At low frequencies
where the dynamic viscosity is independent of ω, the
elasticity as measured by G' is very small. The elastic
modulus G' levels off at a level of about 10^6 or 10^7
dynes/cm^2 at very high frequencies; this modulus value
is comparable to the modulus of a rubber band. From
Figure 3 one sees that the dynamic viscosity η' and the
steady-flow viscosity η are nearly identical at low frequencies or rates of shear. Cox and Merz [29] showed
that the magnitude of the complex viscosity η^* is essentially the same as η_a over the complete frequency or $\dot{\gamma}$
range. This approximate equivalence of η^* and η_a has
been verified for a number of polymers [30-32].

Non-Newtonian behavior develops at large amplitudes
as well as at high frequencies [33]. Steady-state flow
superimposed on the oscillatory motion of a dynamic test
delays the onset of non-Newtonian behavior to higher
frequencies [34]. The reason appears to be because the
steady-state flow destroys some of the molecular entanglements.

V. RHEOLOGY OF EMULSIONS, POLYBLENDS, AND BLOCK POLYMERS

The rheology of emulsions, polyblends, block polymers, and graft polymers is similar in that all of these
are two phase systems in which both phases are fluids.
Emulsions are small drops of one liquid dispersed in another liquid. The polymeric systems, however, may be
considerably more complex in that both phases are sometimes continuous, and the components of the two phases
may be chemically attached to each other. Typical
systems in which both phases are continuous include

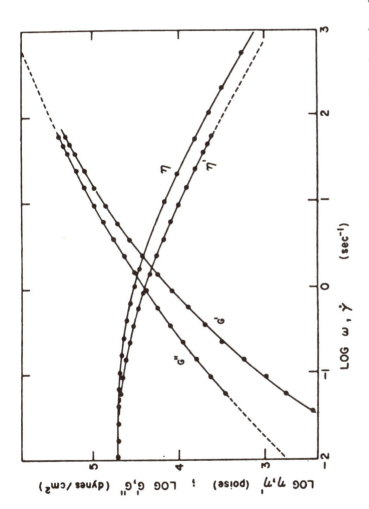

Figure 3.   The apparent viscosity η, the dynamic viscosity η', the elastic shear modulus G', and the dynamic loss factor G" as a function of either the angular frequency ω or the shear rate γ̇. The absolute value of the complex viscosity η* would be nearly identical to η. [Reprinted from Shroff, Trans. Soc. Rheol., 15, 163 (1971).]

interpenetrating networks, structures such as filled
open-celled foams, impregnated fabrics, and lamellar
structures such as found with many block polymers.
Thus, a number of different morphologies are possible
when both phases are continuous.

A. Emulsions

The presence of a sphere in a shear field disturbs
the streamlines so that additional energy is dissipated
by the system. This additional dissipation of energy
manifests itself as an increase in viscosity. Einstein
[35] showed that the viscosity of a dilute suspension of
rigid spheres is

$$\eta = \eta_1 (1 + k_E \phi_2). \qquad (6)$$

The viscosity of the suspending medium is η_1, the volume
fraction of spheres is ϕ_2, and k_E is the Einstein coefficient. The Einstein coefficient is 2.50 for rigid
spheres if there is no slippage of the liquid at the
surface of the sphere. If there is perfect slippage of
liquid at the interface, k_E becomes 1.0 [35-37].

In emulsions there is circulation of fluid within the
spheres in addition to displacement of the streamlines
of the drop [38-40]. The circulation within the drops
allows relative motion to take place in the neighborhood
of the interface in a manner similar to what takes place
during slippage at the interface. Thus, intuitively,
one would expect the Einstein coefficient to be about
2.50 when the viscosity of the fluid in the drops is
much greater than the viscosity of the continuous liquid,
and k_E should be 1.0 when the viscosity of the continuous liquid is much greater than that of the drops.
Taylor [39-41] derived the following equation for the
Einstein coefficient of emulsions at low shear rates:

$$k_E = 2.50 \left(\frac{\eta_2 + 2/5 \ \eta_1}{\eta_2 + \eta_1} \right). \tag{7}$$

The viscosity of the dispersed liquid is η_2. This equation is plotted in Figure 4. It should be pointed out that the drops rotate in a simple shear field at the same time that there is circulation within the drop itself [39,40,42]. The angular rotation rate ω is

$$\omega = \frac{\dot{\gamma}}{2}. \tag{8}$$

The Einstein equation is valid only at very low concentrations. At higher concentrations, many equations have been proposed, but the Mooney equation seems to be as good as any [43]:

$$\eta = \eta_1 \exp \left(\frac{k_E \ \phi_2}{1 - \phi_2 / \phi_m} \right). \tag{9}$$

The maximum packing fraction ϕ_m is the actual volume of the spheres divided by the volume that the spheres appear to occupy. For random close packing, $\phi_m = 0.637$, while for hexagonal close packing, $\phi_m = 0.74$. However, in some emulsions, ϕ_m may approach 1.0 because of the ease with which the spheres can be deformed when they contact one another. Although k_E and ϕ_m have values which can be derived sometimes from theory, the best fit to experimental data often is attained by assigning empirical values to k_E and ϕ_m which are somewhat different from the theoretical values.

Most emulsions contain surfactants which can affect the rheology by at least three mechanisms: 1. The surfactant may form a "skin" around the drops which reduces

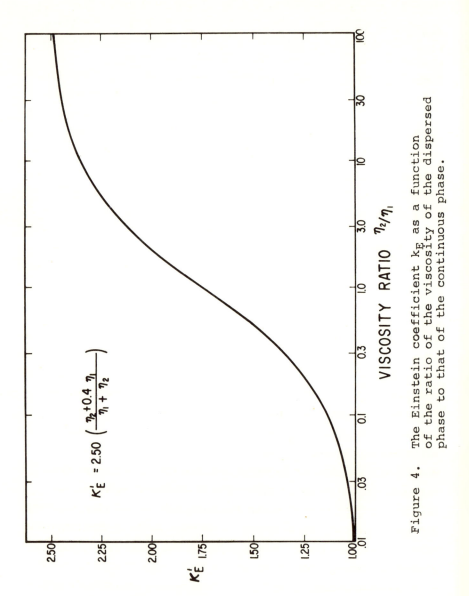

Figure 4.   The Einstein coefficient k_E as a function of the ratio of the viscosity of the dispersed phase to that of the continuous phase.

the amount of circulation within the drops. 2. The surfactant forms a coating on the drops which makes the
drops appear larger than they would be if there were
no surfactant. 3. Some of the surfactant may dissolve
in the suspending fluid and change its viscosity. In
general, all of these factors tend to increase the viscosity of the system.

Interfacial tension keeps the suspended liquid in
the shape of spheres when there is no flow. However,
since liquid drops are easily deformable, the drops may
become distorted from the spherical shape in a shear
field. Thus, the above equations generally are applicable only at low rates of shear. Taylor [44] and Mason
[40,45] have studied the deformation of liquid drops in
flow fields. If the rate of shear $\dot{\gamma}$ is so small as to
distort the drops only slightly, the deformation is
given by [39, 44]:

$$\frac{L - W}{L + W} = \frac{D\eta_1 \dot{\gamma}}{2\mu} \left(\frac{\eta_1 + 1.1875 \, \eta_2}{\eta_1 + \eta_2} \right) \quad (10)$$

The dimensions of the resulting ellipsoid are L and W for
the length of the major and minor axes, respectively.
The diameter of the undistorted spheres is D, and μ is
the interfacial tension. Large drops are more easily
deformed than small ones. The stretching action of the
flow field is counteracted by the interfacial tension,
which tries to restore the ellipsoid to a sphere. In
polymer melts, the elastic modulus may be the major factor in place of interfacial tension in resisting deformation. An approximate theory has been developed which
predicts the viscosity of emulsions when the drops are
slightly deformed by a flow field [46]. As the drops
deform, the viscosity becomes a function of the rate of
shear, and normal stresses develop.

At higher rates of shear, the deformed drops may
either break up into smaller drops or stretch out into
long filaments [45]. Typical types of behavior for Newtonian fluids in a shear field are shown in Figure 5 [45].

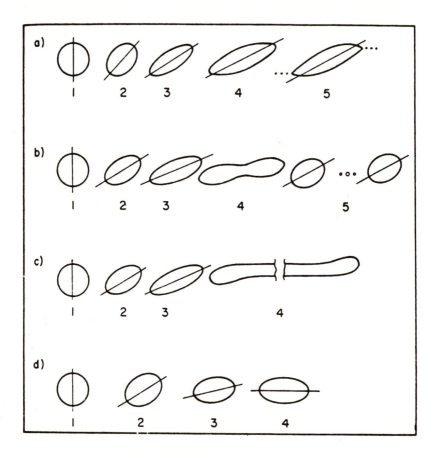

Figure 5.   The shape of drops in shear fields as the
            intensity of the rate of shear increases from
            left to right.   (a) $\eta_2/\eta_1 = 2 \times 10^{-4}$;
            (b) $\eta_2/\eta_1 = 1.0$; (c) $\eta_2/\eta_1 = 0.7$; (d) $\eta_2/\eta_1 =$
            6.0.   [Reprinted from Rumscheidt and Mason,
            J. Colloid Sci., 16, 238 (1961).]

Somewhat similar behavior is observed in elongational
flow fields [45]. As the ratio η_2/η_1 increases, there
is less breakup and more filament formation. Non-
Newtonian polymer melts have a greater tendency to form
filaments than Newtonian liquids do. This is probably
due to entanglements which tend to prevent breakup by
acting as temporary crosslinks.

B. Polyblends, Block, and Graft Polymers

Mixtures of polymers, block polymers, and graft
polymers all have some of the characteristics of emulsions as discussed in the previous section. However,
there are some differences brought about by the following factors: 1. The dispersed phase may not be spherical
in shape. For example, block polymers may have cylindrical domains, or both phases may be continuous [47-49].
2. The interfacial effects characteristic of emulsions
may be completely overwhelmed by elastic effects in
polymer melts. 3. Block and graft polymers have parts
of each molecule in each of the two liquid phases. When
large deformations take place, strands of one kind of
polymer are pulled through the other material. This
phenomenon can greatly increase the viscosity of such
materials.

Probably the simplest polyblend systems rheological-
ly are suspensions of microgel particles in the same or
a different type of polymer as matrix. The microgel
often is produced commercially as a crosslinked rubber
latex used in high impact polymers. The microgel particles increase the viscosity of the matrix as predicted
by the equations of the previous section [50-54]. The
Einstein coefficient increases somewhat as the degree
of crosslinking increases the rigidity of the microgel
particles. Microgel particles greatly reduce the die
swell of a polymer. High impact polystyrenes and ABS

polymers generally contain microgel rubber particles in
which the rubber particles are also graft polymers, and
the morphology of the dispersed particles may be more
complex than simple shears. In these complex two-phase
systems, the dispersed phase increases the viscosity and
decreases the die swell as well as the energy of activation for flow [55,56].

The flow behavior of mixtures of two incompatible
uncrosslinked polymers can be very complex and unpredictable. The dispersed phase need not be spherical in
shape, and its morphology may change with concentration
and rate of shear. There can be an inversion of the
phases, or both polymers can be continuous phases. Thus,
unless one has additional information on the morphology
of the system under all conditions, there is little hope
of making accurate predictions about the viscosity of
such polyblends. A helpful generalization, however, is
that in flowing systems the low viscosity component
tends to become the continuous phase and to encapsulate
the high viscosity component and thus the overall viscosity of the mixture is reduced [57-61]. This effect
is the result of the principle that systems tend to reduce their dissipation of energy to a minimum. Therefore, in capillary flow, the low viscosity component
tends to collect near the high shear region at the wall
while the high viscosity polymer collects in the center
as a core.

At constant temperature and constant shear stress
(or $\dot{\gamma}$), the viscosity of polyblends as a function of
composition may go through a minimum or even both a minimum and a maximum [62,63]. The minimum occurs in a
concentration range where phase inversion is expected to
occur and where both polymers are more or less continuous phases. The first normal stress difference goes

through a maximum at about the same concentration as
where the viscosity goes through a minimum. Without
information of the types in addition to the rheological
data, it is impossible at present to predict such behavior. Probably the best that one can do is to assume
that the logarithmic rule of mixtures applies at constant
temperature and shear rate [61]:

$$\log \eta = \phi_1 \log \eta_1 + \phi_2 \log \eta_2 \qquad (11)$$

In this equation η_1 and η_2 are the viscosities of the
two polymers by themselves at the same temperature, while
ϕ_1 and ϕ_2 are their volume fractions.

As mixtures of two polymers are deformed during
flow, complex morphologies can develop [64-67]. Pieces
of one polymer may be drawn into filaments. These filaments may remain as filaments, they may break up into
small drops, or the filaments may connect to each other
to give an interconnected network. Under some conditions
ribbons or sheets of one polymer can be formed inside of
the other polymer. Thus, it is readily seen that the
rheological properties of polyblends can be very complex.

The viscosity of block and graft polymers is much
greater than what would be expected on the basis of their
total molecular weight and composition except in the
rare cases where the two components are soluble in one
another [68-74]. The high viscosity apparently is a
result of breaking up of the domain structure of the two
phase systems. Filaments of one type of polymer are
pulled through the other polymer during the deformation
and breakup of the domains during flow. Figure 6 illustrates two types of domains found in block polymers.
Spherical and cylindrical morphology exist when one of
the components is less than about thirty percent of the

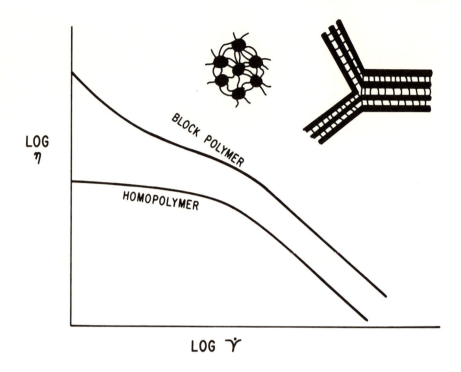

Figure 6.   Comparison of the viscosity of a block
             polymer with a typical homopolymer which has
             the same molecular weight as the total
             molecular weight of the block polymer.
             Inserts show two typical morphologies of
             block polymers with cylindrical (or spherical)
             domains or lamellar domains.

total volume. Lamellar morphology is found when both
components are more equal in volume. The figure also
illustrates how block polymers generally differ from
homopolymers and random copolymers in flow behavior.

VI. REFERENCES

1. J. A. Brydson, Flow Properties of Polymer Melts, Van Nostrand Reinhold, New York, 1970.

2. J. R. Van Wazer, J. W. Lyons, K. Y. Kim, and R. E. Colwell, Viscosity and Flow Measurement, Interscience, New York, 1963.

3. F. T. Wall, J. Chem. Phys., 11, 67 (1943).

4. H. M. James and E. Guth, J. Chem. Phys., 11, 470 (1943).

5. P. J. Flory, Principles of Polymer Chemistry, Cornell Univ. Press, Ithaca, N. Y., 1953.

6. L. R. G. Treloar, Physics of Rubber Elasticity, 2nd Ed., Clarendon Press, Oxford, 1958.

7. L. R. G. Treloar, Trans. Faraday Soc., 36, 538 (1940).

8. R. Buchdahl, J. Colloid Sci., 3, 87 (1948).

9. L. E. Nielsen and R. Buchdahl, J. Chem. Phys., 17, 839 (1949).

10. L. E. Nielsen and R. Buchdahl, J. Colloid Sci., 5, 282 (1950).

11. F. Bueche, J. Chem. Phys., 25, 599 (1956).

12. W. W. Graessley, Entanglement Concept in Polymer Rheology, Adv. Polymer Sci., Vol. 16, Springer-Verlag, Berlin, 1974.

13. R. Buchdahl, J. Colloid Sci., 3, 87 (1948).

14. E. R. Howells and J. J. Benbow, Trans. Plastics Inst. 30, 240 (1962).

15. J. H. Prickard and K. F. Wissbrun, J. Appl. Polymer Sci., 13, 233 (1969).

16. F. Bueche, J. Chem. Phys., 48, 4781 (1968).

17. F. Bueche and S. Harding, J. Polymer Sci., 32, 177 (1958).

18. F. Bueche, Physical Properties of Polymers, Interscience, New York, 1962.

19. W. W. Graessley, J. Chem. Phys., 47, 1942 (1967).

20. W. W. Graessley, R. L. Hazleton, and L. R. Lindeman, Trans. Soc. Rheol., 11, 267 (1967).

21. M. C. Williams, A.I.Ch.E. J., 12, 1064 (1966).

22. M. C. Williams, A.I.Ch.E. J., 13, 534 (1967).

23. M. M. Cross, J. Appl. Polymer Sci., 13, 765 (1969).

24. H. Schott, Rheol. Acta, 7, 179 (1968).

25. J. D. Ferry, Viscoelastic Properties of Polymers, 2nd Ed., John Wiley, New York, 1970.

26. L. E. Nielsen, Mechanical Properties of Polymers and Composites, Vol. 1, Marcel Dekker, New York, 1974.

27. R. N. Shroff, Trans. Soc. Rheol., 15, 163 (1971).

28. T. W. Huseby and L. L. Blyler, Jr., Trans. Soc. Rheol., 11, 77 (1967).

29. W. P. Cox and E. H. Merz, J. Polymer Sci., 28, 619 (1958).

30. S. Onogi, T. Fujii, H. Kato, and S. Ogihara, J. Phys. Chem., 68, 1598 (1964).

31. S. Onogi, T. Masuda, and T. Ibaragi, Kolloid Zeit., 222, 110 (1968).

32. D. W. Verser and B. Maxwell, Polymer Eng. Sci., 10, 122 (1970).

33. T-T. Tee and J. M. Dealy, Trans. Soc. Rheol., 19, 595 (1975).

34. J. M. Simmons, Rheol. Acta, 7, 184 (1968).

35. A. Einstein, Ann. Physik, 19, 289 (1906); 34, 591 (1914).

36. H. L. Frisch and R. Simha, Rheology, Vol. 1, F. R. Eirich, Ed., Academic Press, New York, 1956, p. 525.

37. J. Happel and H. Brenner, Low Reynolds Number Hydrodynamics, Prentice-Hall, New York, 1965.

38. W. Bartok and S. G. Mason, J. Colloid Sci., 13, 293 (1958).

39. P. Sherman, Emulsion Science, Academic Press, New York, 1968.

40. H. L. Goldsmith and S. G. Mason, Rheology, Vol. 4 F. R. Eirich, Ed., Academic Press, New York, 1967, p. 85.

41. G. I. Taylor, Proc. Royal Soc., A138, 41 (1932).

42. G. B. Jeffery, Proc. Royal Soc., A102, 161 (1922).

43. M. Mooney, J. Colloid Sci., 6, 162 (1951).

44. G. I. Taylor, Proc. Royal Soc., A146, 501 (1934).

45. F. D. Rumscheidt and S. G. Mason, J. Colloid Sci., 16, 238 (1961).

46. W. R. Schowalter, C. E. Chaffey, and H. Brenner, J. Colloid Interf. Sci., 26, 152 (1968).

47. T. Inoue, T. Soen, T. Hashimoto, and H. Kawai, J. Polymer Sci., A2, 7, 1283 (1969).

48. T. Inoue, T. Soen, T. Hashimoto, and H. Kawai, Macromol., 3, 87 (1970).

49. A.Douy and B.Gallot, Makromol.Chem., 156, 81 (1972).

50. S. L. Rosen and F. Rodriguez, J. Appl. Polymer Sci., 9, 1601, 1615 (1965).

51. G. V. Vinogradov, V. N. Kuleznev, A. Ya. Malkin, A. V. Igumnova, and O. G. Polyakov, Colloid J. (USSR), 29, 145 (1967), (English trans.)

52. A. Casale, A. Moroni, and C. Spreafico, Adv. Chem. Series (Am. Chem. Soc.), 142, 172 (1975).

53. S. Newman and Q. Trementozzi, J. Appl. Polymer Sci., 9, 3071 (1965).

54. R. S. Hagan and D. A. Davis, J. Polymer Sci., B2, 909 (1964).

55. H. Kubota, J. Appl. Polymer Sci., 19, 2299 (1975).

56. R. L. Bergen, Jr. and H. L. Morris, Proc. 5th Internat. Congr. Rheol., Vol. 4, p. 433, S. Onogi, Ed., Univ. Tokyo Press, Tokyo, 1970.

57. J. H. Southern and R. L. Ballman, Appl. Polymer Symposium, No. 20, 175 (1973).

58. J. L. Work, Polymer Eng. Sci., 13, 46 (1973).

59. A. E. Everage, Jr., Trans.Soc.Rheol., 17, 629 (1973).

60. J. H. Southern and R. L. Ballman, J. Polymer Sci., (Phys.), 13, 863 (1975).

61. B-L. Lee and J. L. White, Trans. Soc. Rheol., 19, 481 (1975).

62. C. D. Han, Y. W. Kim, and S. J. Chen, J. Appl. Polymer Sci., 19, 2831 (1975).

63. C. D. Han and Y. W. Kim, Trans. Soc. Rheol., 19, 245 (1975).

64. H. Vanoene, J. Colloid Interf. Sci., 40, 448 (1972).

65. J. M. Starita, Trans. Soc. Rheol., 16, 339 (1972).

66. V. L. Folt and R. W. Smith, Rubber Chem. Techn., 46, 1193 (1973).

67. M. V. Tsebrenko, M. Jakob, M. Yu. Kuchinka, A. V. Yudin, and G. V. Vinogradov, Internat. J. Polymer.Mater., 3, 99 (1974).

68. G. Kraus and J. T. Gruver, J. Appl. Polymer Sci., 11, 2121 (1967).

69. G. Holden, E. T. Bishop, and N. R. Legge, J. Polymer Sci., C26, 37 (1969).

70. G. M. Estes, S. L. Cooper, and A. V. Tobolsky, J. Macromol. Chem., C4, 313 (1970).

71. K. R. Arnold and D. J. Meier, J. Appl. Polymer Sci., 14, 427 (1970).

72. F. N. Cogswell and D. E. Hanson, Polymer, 16, 936 (1975).

73. C. I. Chung and J. C. Gale, J. Polymer Sci. (Phys). 14, 1149 (1976).

74. N. V. Chii, A. I. Isayev, A. Ya. Malkin, G. V. Vinogradov, and I. Yu. Kirchevskaya, Polymer Sci. USSR, 17, 983 (1975). (English trans.)

Chapter 5

EFFECTS OF MOLECULAR WEIGHT AND STRUCTURE

I.	MOLECULAR WEIGHT DEPENDENCE OF VISCOSITY	69
II.	EFFECTS OF DISTRIBUTION OF MOLECULAR WEIGHTS ON VISCOSITY	72
III.	DEPENDENCE OF DYNAMIC MECHANICAL PROPERTIES ON MOLECULAR WEIGHT	75
IV.	EFFECTS OF STRUCTURE ON POLYMER RHEOLOGY	80
	A. Branching	80
	B. Other Structural Factors	82
V.	REFERENCES	83

I. MOLECULAR WEIGHT DEPENDENCE OF VISCOSITY

Among the structural factors determining the rheology of a polymer, the molecular weight is the most important [1-7]. Below a critical molecular weight Me, the viscosity of a molten polymer is roughly proportional to the weight average molecular weight $\overline{M}w$. That is,

$$\eta \doteq K_1 \ \overline{M}w \qquad\qquad \overline{M}w \leq Me \qquad\qquad (1)$$

At molecular weights above Me, the viscosity at low rates of shear depends upon $\overline{M}w$ to a power equal to about 3.4 or 3.5.

$$\eta_\odot \doteq K_2 \ \overline{M}w^{3 \cdot 4} \qquad\qquad \overline{M}w \geq Me \qquad\qquad (2)$$

Me is the molecular weight at which chain entanglements become important. Once the chains are long enough to become entangled, flow becomes much more difficult

69

because forces applied to one polymer chain become
transmitted to and distributed among many other chains.
For most polymers, Me is between about 10,000 and 40,000.
Some typical values are given in Table 1.

Table 1
Entanglement Molecular Weight Me

Polymer	Me
Linear PE	4000
Polyisobutylene	17,000
Polyvinyl Acetate	29,200
Polystyrene	38,000
Polydimethyl siloxane	35,200
Polymethyl methacrylate	10,400
Caprolactam polymer (linear)	19,200
Caprolactam polymer (tetra-branched)	22,000
Caprolactam polymer (octa-branched)	31,100

The constants K_1 and K_2 depend upon temperature in the
manner already discussed in Chapter 3. These constants
also depend upon the molecular structure; the constants
are smaller for flexible molecules such as the dimethyl
silicones than for more rigid molecules such as highly
condensed aromatic ring polymers.

The complete molecular weight dependence of the
viscosity is illustrated in Figure 1. The solid lines
show the dependence of the viscosity on molecular weight
at essentially zero rate of shear. There is some doubt
about this dependence at higher shear stresses as indicated by the dashed lines in Figures 1A and 1B. The
dashed lines in Figures 1A and 1B indicate two of the

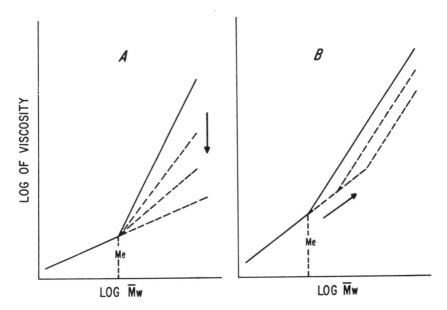

Figure 1.   Two possible ways in which the melt viscosity
            varies with molecular weight. Solid lines
            denote η_0. Dashed lines are for higher shear
            stresses, which increase in the direction of
            the arrows.

possibilities with the arrows indicating the direction
of the shifts as τ increases. In Figure 1A the slope of
the line gradually decreases from the 3.4 power dependence at zero rate of shear to about a linear dependence
on molecular weight at very high shear stresses. This
type of η-$\overline{M}w$ dependence might be expected if shearing
destroys the entanglements faster than they can reform
so that the number of entanglements decreases as $\dot{\gamma}$ increases. The type of behavior illustrated in Figure 1B
would be expected if Me increases as $\dot{\gamma}$ or τ increases.
The slope would remain at 3.4 at molecular weights above
Me, but Me would shift with the rate of shear [5]. The
curves shown in Figure 1A might be found if the viscosity of different molecular weight fractions is compared

at the same shear stress τ. If the fractions are compared at the same rate of shear $\dot{\gamma}$ instead of at the same
shear stress, the straight lines may become curved and
reach asymptotic values of viscosity at very high molecular weights [8]. In the extreme case, the viscosity at
molecular weights above Me may become nearly independent
of $\overline{M}w$ at very high rates of shear. It is difficult to
decide between the behaviors illustrated in Figures 1A
and 1B unless a very large number of different molecular
weight fractions are available for a detailed study over
a wide range of shear rates or shear stresses.

Typical viscosity versus shear rate curves are
shown in Figure 2 for monodisperse polystyrenes of different molecular weights [5]. Since low molecular
weight polymers have fewer entanglements than high molecular weight ones, it is not surprising that deviations
from Newtonian behavior start at higher shear rates for
the low molecular weight materials [5,9].

Master curves can be produced by both horizontal
and vertical shifts of curves such as those in Figure 2.
A single master curve generally can be produced for all
monodisperse molecular weights, all shear rates, and all
temperatures by plotting log (η/η_o) against log
$(\dot{\gamma}\eta_o M_w^{\alpha}/\rho T)$. The constant α is generally near 1.0, but
it may vary somewhat. The temperature T is in degrees
Kelvin, and ρ is the density of the polymer at each
temperature [4,5,10-13].

II. EFFECTS OF DISTRIBUTION OF MOLECULAR WEIGHTS ON
                    VISCOSITY

The distribution of molecular weights in a polymer
has an effect on its rheology. For fractions and polymers with a narrow distribution of molecular weights,
the weight average molecular weight is the important

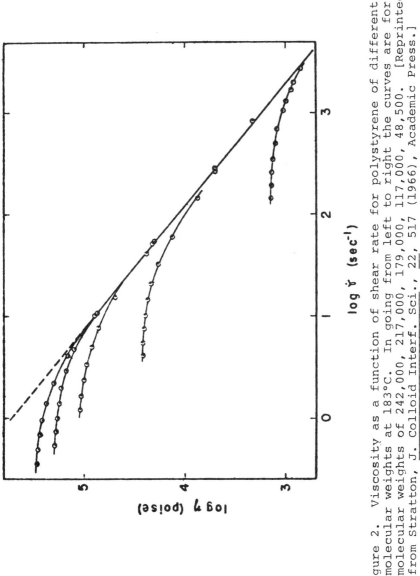

Figure 2. Viscosity as a function of shear rate for polystyrene of different molecular weights at 183°C. In going from left to right the curves are for molecular weights of 242,000, 217,000, 179,000, 117,000, 48,500. [Reprinted from Stratton, J. Colloid Interf. Sci., 22, 517 (1966), Academic Press.]

molecular weight [10,14-16). For blends of two fractions
of different molecular weights of a polymer, η_o appears
to be uniquely determined by $\overline{M}w$ also [13].

For normal whole polymers and other polymers with a
wide distribution of molecular weights, the melt viscosity may not be strictly dependent upon $\overline{M}w$ [16,17].
For such polymers, the log η_o is proportional to an
average molecular weight which is greater than $\overline{M}w$. The
proper molecular weight may be some average which is
between $\overline{M}w$ and $\overline{M}z$, which is the next higher moment of
the distribution curve beyond $\overline{M}w$. Thus, the high molecular weight "tails" on the distribution curve can be
especially important in affecting η_o and other rheological properties. Although $\overline{M}w$ dominates the viscosity at
low rates of shear, a molecular weight between $\overline{M}n$ and $\overline{M}w$
appears to determine the viscosity at high rates of
shear [10,14]. However, the conflicting effects of pressure and viscous heating on the measured apparent viscosity may make the molecular weight dependence of viscosity
ambiguous at high rates of shear.

The distribution in molecular weight affects the
value of the shear rate at which non-Newtonian behavior
becomes apparent. The polymer with a broad distribution
exhibits non-Newtonian flow at a lower rate of shear than
a polymer with the same zero rate of shear viscosity but
which has a narrow distribution in molecular weights [10,
18,19]. The width of the distribution curve can be ex-
pressed by $\overline{M}w/\overline{M}n$, which increases as the distribution
broadens. The shift in non-Newtonian behavior to lower
$\dot{\gamma}$ as the distribution broadens has important practical
implications. Fractions or polymers with a very narrow
distribution appear to have a higher viscosity than the
same polymer with a normal or broad distribution under
conventional molding conditions where the rate of shear

is high [20,21]. Thus, polymers with a normal or broad
distribution of molecular weights are generally easier
to extrude or mold than a polymer with a narrow distribution.

III. DEPENDENCE OF DYNAMIC MECHANICAL PROPERTIES ON
            MOLECULAR WEIGHT

Typical dynamic mechanical properties of polymer
melts as a function of frequency at different molecular
weights are shown in Figures 3 and 4 [6]. The molecular
weights of the polymers are given in Table 2. Of course,
the dynamic viscosity η' can be calculated from the loss
modulus G'' by

$$\eta' = G''/\omega. \tag{3}$$

A typical example of log η' versus frequency curves is
shown in Figure 5 [22]. Both the loss modulus and the
elastic modulus increase rapidly with frequency and with
molecular weight up to a plateau value of about 10^6 to
10^7 dynes/cm^2. Low molecular weight polymers which have
few if any entanglements have little or no plateau
region in the log G' or log G'' versus log ω plots. High
molecular weight polymers have plateau regions which
cover several decades in frequency. These plateaus exist in a frequency range above which the entanglements
do not have time to slip and relax out the stress.
These values of G' and G'' in the plateau regions are
essentially independent of molecular weight because the
effective molecular weight between entanglements becomes
constant. Entanglements which can not relax out behave
as crosslinks. Thus, from the kinetic theory of rubber
elasticity [23],

$$G' \doteq \frac{\rho RT}{Me}. \tag{4}$$

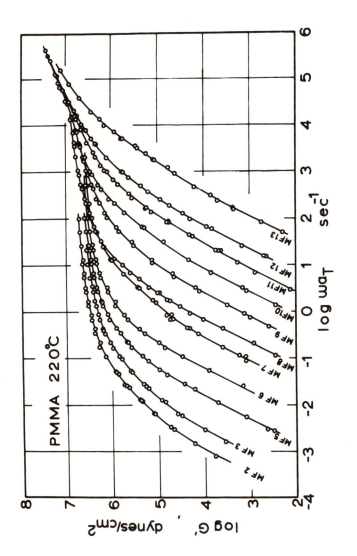

Figure 3. Dynamic modulus master curves for polymethyl methacrylates of different molecular weights. Reference temperature is 220°C. Molecular weights are listed in Table 2. [Reprinted from Masuda, et al., Polymer J., 1, 418 (1970).]

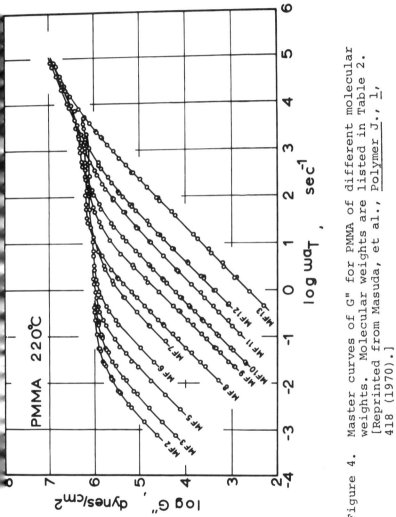

Figure 4. Master curves of G" for PMMA of different molecular
weights. Molecular weights are listed in Table 2.
[Reprinted from Masuda, et al., Polymer J., 1,
418 (1970).]

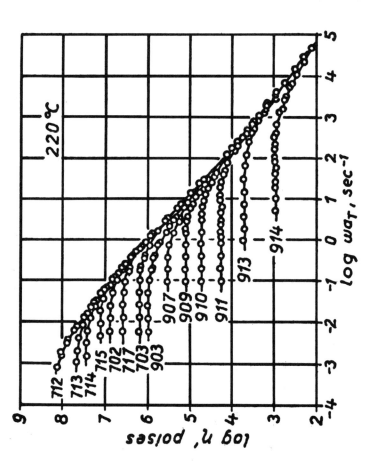

Figure 5.   Master curves of η' for the PMMA fractions listed in
Table 2. [Reprinted from Onogi, et al., Koll. Zeit.
222, 110 (1968) with permission of Dr. Dietrich
Steinkopff Verlag, Darmstadt, W. Germany.]

Table 2

Molecular Weights of Polymethyl Methacrylate Fractions
Used in Figures 3,4, and 5

Polymer	Mw	Mn
MF2	342,000	224,000
MF3	270,000	188,000
MF5	197,000	146,000
MF6	158,000	139,000
MF7	116,000	98,000
MF8	96,300	71,800
MF9	63,900	52,800
MF10	45,200	37,100
MF11	35,100	25,300
MF12	28,400	18,000
MF13	11,400	9,120
712	408,000	251,000
713	278,000	203,000
714	229,000	175,000
715	191,000	151,000
702	174,000	139,000
717	143,000	116,000
703	114,000	99,000
903	101,000	50,000
907	80,100	50,300
909	65,800	43,100
910	52,100	37,500
911	40,800	32,800
913	29,100	21,000
914	18,900	9,270

The use of this equation is one way of calculating Me.
However, the molecular weight between entanglements Me
calculated by this equation is roughly half that obtained
from a plot of log η_o versus log Mw. A possible reason
for this difference is that the lowest molecular weight
which can form an entanglement must be about twice the
molecular weight between entanglements of a much longer
chain because about half the total length of a short chain
must be on each side of the point of entanglement with
another chain.

All the dynamic properties G', G" and η' increase
in value with molecular weight increase at a given frequency. Non-Newtonian behavior in η' becomes observable
when the dynamic shear modulus reaches a value of about
10^5 dynes/cm^2. Also, in the frequency range where the G
values converge into a plateau, the viscosity curves all
converge into a single curve as shown in Figure 5 [22].

The upturn in the curves in Figures 3 and 4 at
reduced frequencies greater than 10^4 radians/second is
due to the beginning of the glass transition region. The
polymer melt no longer behaves as a fluid but begins to
take on the characteristics of a glass at these high frequencies, and many kinds of molecular motions are no
longer able to take place in time intervals less than
10^{-4} seconds. Dynamic mechanical behavior similar to
that shown in Figures 3 to 5 has been observed for a
number of other polymers [24-28]. Master curves can be
produced by plotting $\ln(\eta'/\eta_o)$ or $\ln(\eta''/\eta_o)$ against
$\ln(\omega\eta_o Mw/\rho T)$.

IV. EFFECTS OF STRUCTURE ON POLYMER RHEOLOGY
A. Branching

Polymers can have a great variety of branched
structures. The branches can be long or short. The

branches can be randomly spaced along the backbone
chain, or several branches can originate from a single
point to give a star-shaped molecule.

Short branches generally do not affect the viscosity of a molten polymer very much. On the other hand,
long branches can have a very large effect. Branches
which are long but which are still shorter than those
required for entanglements decrease the viscosity when
compared to a linear polymer of the same molecular
weight [29-31]. Such branching reduces the viscosity
because branched molecules are more compact than linear
molecules. Thus, there tends to be less intermolecular
interaction between branched molecules. However, if the
branches are so long that they can participate in entanglements, the branched polymer may have a viscosity
at low rates of shear greater than that of a linear polymer of the same molecular weight [32-34]. However, in
other cases, the viscosity at low rates of shear may be
less for branched polymers than for linear polymers of
the same molecular weight even though the branches appear
to be long enough to form entanglements [25,29,30,31,
35]. The reason why long branches appear to increase η_o
in some cases and to decrease η_o in others is not clear.
Possibly, the entanglement molecular weight Me is considerably greater for branched than for linear chains
because of the compact nature of branched chains [25,31].
Also, branched molecules often have a higher shear rate
dependence of viscosity than linear molecules, so in
some experimental work the shear rates may have been too
high to achieve the highest possible viscosity [33].
Thus, at high rates of shear, branched polymers in
nearly all cases have lower viscosity than linear ones
of the same molecular weight. Although, increasing the

number of branches tends to decrease the viscosity, the
molecular weight of the branches seems to be more important than their number [32]. Long branches also increas
the energy of activation for flow [30,36,37].

Long branches generally decrease the elasticity or
shear modulus G' of polymer melts [25,29,31]. This is
somewhat surprising as one would intuitively expect
branches to increase the shear modulus if the branches
are long enough to form strong entanglements.

The manufacturer of polymers thus has means of controlling the viscosity, the elasticity, and the shear
rate dependence of viscosity and elasticity by variation
in molecular weight and length of branches. These rheological factors in turn affect the processing or fabrication and the mechanical characteristics of finished
objects [30,35,38].

B.   Other Structural Factors

Flexible chains have lower viscosities than
rigid polymers of comparable molecular weight. Silicone$
and polymers containing ether linkages have especially
low viscosities [39]. Very rigid polymers, such as poly-
imides and other condensed aromatic ring polymers, have
very high viscosities and are difficult to process.

Any factor which increases the glass transition
temperature tends to increase viscosity. These factors
include polarity, hydrogen bonding, and ionic bonding,
in addition to chain stiffness [40]. Hydrogen bonding
may moderately increase the viscosity of nylons, polyvinyl alcohol, and polyacrylic acid [41,42]. However,
ionic bonding, such as found in polyelectrolytes, greatl$
increases the viscosity [43,44]. Ionic bonds may tie
molecules together nearly as strongly as crosslinks.

Some polar polymers, such as polyvinyl chloride and
polyacrylonitrile, have strong intermolecular bonding an$

may actually be slightly crystalline even in the molten state. These polymers have very high viscosities and elastic moduli. Polymers such as polyvinyl chloride can have vastly different rheological behavior depending upon the way in which they are polymerized. Mass and suspension polymerized materials do not show the same rheological properties as emulsion polymerized ones [45-47]. The emulsion particles of polyvinyl chloride act as relatively rigid flow units which can slip by one another with little interaction between the molecules making up each individual emulsion particle. Emulsion polymerized polyvinyl chloride has lower viscosity, less die swell, and smoother extrudate than mass polymerized polymer. At high temperatures (above 200°C) the flow units of emulsion particles disappear, the energy of activation for flow increases, and the polymer then behaves normally.

Thermosetting polymers, such as phenol-formaldehyde resins and epoxy resins, initially have a low viscosity, but the viscosity rapidly increases as the polymerization reaction proceeds. The crosslinking reactions eventually produce a gel, at which point the viscosity and elastic modulus greatly increase in a very short time [48,49]. After the gel point, the viscosity increases so rapidly that soon no flow is possible.

V. REFERENCES

1. T. G. Fox and P. J. Flory, J. Amer. Chem. Soc., 70, 2384 (1948).

2. T. G. Fox and P. J. Flory, J. Polymer Sci., 14, 315 (1954).

3. T. G. Fox, S. Gratch, and S. Loshaek, Rheology, Vol. 1, F. R. Eirich, Ed., Academic Press, New York, 1956, p. 431.

4. A. Casale, R. S. Porter, and J. F. Johnson,
    J. Macromol. Sci., C5, 387 (1971).

5. R. A. Stratton, J. Colloid Interf. Sci., 22, 517
    (1966).

6. T. Masuda, K. Kitagawa, and S. Onogi, Polymer J.,
    1, 418 (1970).

7. F. Bueche, J. Appl. Phys., 24, 423 (1953); 26, 738
    (1955).

8. A. Casale, A. Moroni, and E. Civardi, Ang. Makromol
    Chem., 53, 1 (1976).

9. H. P. Schreiber, E. B. Bagley, and D. C. West,
    Polymer, 4, 355 (1963).

10. R. L. Ballman and R. H. M. Simon, J. Polymer Sci.,
    A2, 3557 (1964).

11. R. L. Ballman, Nature, 202, 288 (1964).

12. W. W. Graessley, R. L. Hazelton, and L. R. Lindeman
    Trans. Soc. Rheol., 11, 267 (1967).

13. W. W. Graessley, Adv. Polymer Sci., Vol. 16,
    Springer Verlag, Berlin, 1974.

14. J. F. Rudd, J. Polymer Sci., 44, 459 (1960).

15. E. M. Friedman and R. S. Porter, Trans. Soc. Rheol.
    19, 493 (1975).

16. A. Rudin and K. K. Chee, Macromol., 6, 613 (1973).

17. V. R. Allen and T. G. Fox, J. Chem. Phys., 41, 337
    (1964).

18. R. Sabia, J. Appl. Polymer Sci., 8, 1053 (1964).

19. C. K. Shih, Trans. Soc. Rheol., 14, 83 (1970).

20. H. L. Wagner and K. F. Wissbrun, SPE Trans., 2,
    222 (1962).

21. G. Kraus and J. T. Gruver, J. Appl. Polymer Sci., 9
    739 (1965).

22. S. Onogi, T. Masuda, and T. Ibaragi, Kolloid Zeit,
    222, 110 (1968).

23. P. J. Flory, Principles of Polymer Chemistry, Cornell University Press, Ithaca, N. Y., 1953.

24. S. Onogi, T. Masuda, and K. Kitagawa, Macromol., 3, 109, 116 (1970).

25. T. Masuda, Y. Ohta, and S. Onogi, Macromol., 4, 763 (1971).

26. W. P. Cox, L. E. Nielsen, and R. Keeney, J. Polymer Sci., 26, 365 (1957).

27. M. L. Williams and J. D. Ferry, J. Colloid Sci., 9, 479 (1954).

28. R. N. Shroff, J. Appl. Phys., 41, 3652 (1970).

29. R. A. Mendelson, Polymer Eng. Sci., 9, 350 (1969).

30. J. Miltz and A. Ram, Polymer Eng. Sci., 13, 273 (1973).

31. L. A. Utracki and J. E. L. Roovers, Macromol., 6, 373 (1973).

32. V. C. Long, G. C. Berry, and L. M. Hobbs, Polymer, 5, 517 (1964).

33. G. Kraus and J. T. Gruver, J. Polymer Sci., A3, 105 (1965).

34. W. W. Graessley, T. Masuda, J. E. L. Roovers, and N. Hadjichristidis, Macromol., 9, 127 (1976).

35. J. P. Hogan, C. T. Levett, and R. T. Werkman, SPE J., 23, #11, 87 (1967).

36. L. Boghetich and R. F. Kratz, Trans. Soc. Rheol., 9, 255 (1965).

37. R. A. Mendelson, Trans. Soc. Rheol., 9, 53 (1965).

38. R. N. Shroff, L. V. Cancio, and M. Shida, Modern Plastics, 52, #12, 62 (1975).

39. E. Schonfeld, J. Polymer Sci., 2A, 2489 (1964).

40. L. E. Nielsen, Mechanical Properties of Polymers and Composites, Vol. I, Marcel Dekker, New York, 1974.

41. J. K. Rieke and G. M. Hart, J. Polymer Sci., 1C, 117 (1963).

42. L. L. Blyler, Jr. and T. W. Haas, J. Appl. Polymer Sci., 13, 2721 (1969).

43. W. E. Fitzgerald and L. E. Nielsen, Proc. Royal Soc., A282, 137 (1964).

44. J. Economy, J. H. Mason, and L. C. Wohrer, J. Polymer Sci., A1, 8, 2231 (1970).

45. A. R. Berens and V. L. Folt, Trans. Soc. Rheol., 11 95 (1967).

46. A. R. Berens and V. L. Folt, Polymer Eng. Sci., 8, 5 (1968); 9, 27 (1969).

47. E. A. Collins and C. A. Krier, Trans. Soc. Rheol., 11, 225 (1967).

48. M. B. Roller, Polymer Eng. Sci., 15, 406 (1975).

49. F. G. Massati and C. W. Macosko, Polymer Eng. Sci., 13, 236 (1973).

Chapter 6

EFFECTS OF SOLVENTS, PLASTICIZERS AND LUBRICANTS

I.   RHEOLOGY OF POLYMERS CONTAINING SOLVENTS   87
     AND PLASTICIZERS

     A.   Introduction   87

     B.   Viscosity as a Function of   89
          Concentration of Solvent

     C.   Temperature and Shear Rate Dependence   94
          of the Viscosity of Polymer Solutions

     D.   Dynamic Rheological Properties of   96
          Solutions

II. LUBRICANTS   99

III. REFERENCES   101

I.   RHEOLOGY OF POLYMERS CONTAINING SOLVENTS
                AND PLATICIZERS

A.   Introduction

There are many instances where polymers contain
solvents or plasticizing liquids. The liquids may have
a concentration of only a percent or so, or the liquids
may be greater in volume than the polymer itself. In
some cases the liquids are added for a purpose such as
to plasticize the polymer, to improve its processibility,
or to stabilize the polymer to processing conditions.
In other cases the liquid may be an unwanted impurity
such as a small amount of monomer left over from the
polymerization process.

A major effect of solvents and plasticizers is the
lowering of the glass transition temperature T_g of a
polymer [1]. Liquids also have glass transitions,
generally well below 0°C and in some cases far below

-100°C. When a polymer is mixed with a solvent, the
mixture has a T_g between the values of the two components. A useful mixture rule is

$$T_g = T_{gP}\phi_P + T_{gL}\phi_L - I\phi_P\phi_L \tag{1}$$

where T_{gP} is the glass transition temperature of the
polymer, and T_{gL} is the T_g of the liquid, ϕ_P is the
volume fraction of the polymer, and I is an interaction
term. Often the glass transition temperature of a liquid is about 2/3 of its melting point in degrees Kelvin.
The interaction term must be calculated from the T_g of a
mixture; it is generally small. If only the values for
the pure components are available, the best guess is to
assume that I = 0. Another mixture rule has been derived theoretically [2,3]:

$$T_g = \frac{T_{gP} + (KT_{gL} - T_{gP})\phi_L}{1 + (K - 1)\phi_L} \tag{2}$$

$$K \doteq \frac{\alpha_L}{\alpha_P} \tag{3}$$

The difference between volume coefficients of expansion
above and below the glass transition temperature of the
solvent is α_L. The corresponding difference in coefficients of expansion for the polymer is α_P. Generally,
K is between 1.0 and 2.0. When K = 1, equation 2 is the
same as equation 1 with I = 0. As already indicated in
Chapter 3, the viscosity of a polymeric material is very
dependent upon $(T-T_g)$ where T is the temperature used to
measure the viscosity.

The viscosity of a polymer containing some solvent
is influenced by at least three factors: 1. The solvent
lowers the T_g of the polymer. The effect of T_g on

viscosity was discussed in Chapter 3.   The W-L-F equation
at least approximately accounts for the decrease in viscosity as a result of lowering T_g [4].   2. The liquid
increases the molecular weight M_e of the polymer between
entanglement points [2,5]. Although there is some dis-
agreement on just how M_e varies with ϕ_p, as a first
approximation, the molecular weight between entanglements
can be expressed as

$$M_e = M_e^o/\phi_p \qquad (4)$$

where M_e^o is the value for the pure polymer.   As discussed in chapter 5, the viscosity is very dependent
upon M_e. 3. The viscosity of a high viscosity liquid is
lowered by a liquid of low viscosity because of dilution
according to some "rule of mixtures" even if the first
two of the above factors are not applicable.

B.   Viscosity as a Function of Concentration of
Solvent

There is no completely satisfactory theory for describing the viscosity of a polymer as a function of the
concentration of an added solvent, although a number of
attempts have been made.   In the extreme case of a low
viscosity liquid added to a polymer near its glass transition, the viscosity may change by a factor of 10^{15} in
covering the concentration range from the pure polymer
to the pure liquid.   Even at temperatures well above T_g,
the viscosity can easily vary over a range of a million
to one in going from the polymer to the liquid.   This
range is an extremely wide one to expect any theory to
be valid.   Experimentally, the viscosity of a polymer
solution has been found to vary from the first power to
greater than the fourteenth power of the concentration

of the polymer, depending upon the concentration range, the polymer-solvent system, temperature, and other experimental conditions [6]. Typical experimental data are shown in Figure 1 [7].

Generally, the simplest method of crudely estimating the viscosity of a solution at a temperature well above T_g is to use a mixture rule such as

$$\log \eta \doteq \phi_P \log \eta_P + \phi_L \log \eta_L . \tag{5}$$

The viscosities of the pure polymer and liquid are η_P and η_L, respectively, while ϕ_P and ϕ_L are the corresponding volume fractions. Various theories have been proposed, but their predictive powers often are no better than equation 5, and the theoretical equations are much more difficult to use. Theories about the viscosity of concentrated polymer solutions have been proposed by Kelley and Bueche [8], Fujita and Kishimoto [7,9], Kraus and Gruver [10], and Lyons and Tobolsky [11].

The Kelley-Bueche [8] theory appears to be the best for predicting the viscosity of concentrated polymer solutions using the viscosity of the pure polymer η_P as the reference [12,13]. This theory is based upon the concept that the free volume determines the viscosity. The theory combines the W-L-F equation 4 with an equation of Cohen and Turnbull [14]. The resulting equations are:

$$\frac{\eta}{\eta_P} = \phi_P^4 \exp \left[\frac{1}{\phi_P f_P + \phi_L f_L} - \frac{1}{f_P} \right] \tag{6}$$

$$f_P \doteq 0.025 + 4.8 \times 10^{-4} (T - T_{gP}) \tag{7}$$

$$f_L \doteq 0.025 + \alpha_L (T - T_{gL}) \tag{8}$$

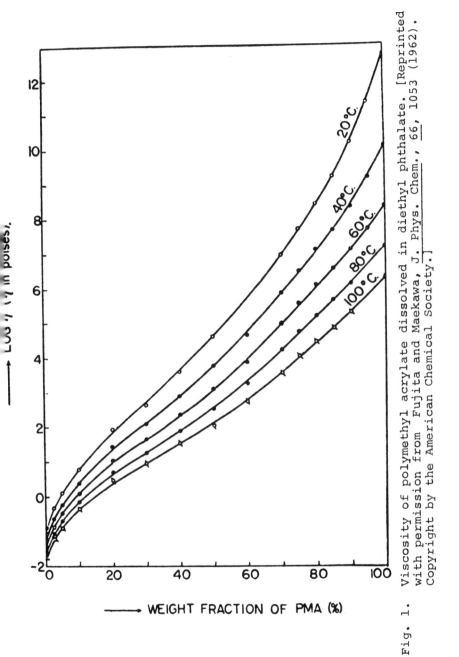

Fig. 1. Viscosity of polymethyl acrylate dissolved in diethyl phthalate. [Reprinted with permission from Fujita and Maekawa, J. Phys. Chem., 66, 1053 (1962). Copyright by the American Chemical Society.]

The free volumes of the polymer and the liquid are f_P and f_L, respectively. The difference in the coefficients of expansion of the liquid above and below T_g is α_L. Generally, $\alpha_L \doteq 10^{-3}$. The glass transition temperatures of the polymer and solvent are T_{gP} and T_{gL}, respectively. The corresponding volume fractions are ϕ_P and ϕ_L. Figure 2 shows plots of equation 6 for several values of f_P and f_L. The figure shows the tremendous sensitivity of the relative viscosity to small changes in f_P and f_L. The concentration dependence of relative viscosity is especially great at temperatures close to the T_g of the pure polymer as represented by curve D. Very small amounts of a liquid or plasticizer can reduce the viscosity of the polymer by one or more decades. Equations 7 and 8 give approximate theoretical values of f_P and f_L. However, because of the extreme sensitivity of viscosity to the free volumes, in practical situations it may be more convenient to consider f_P and f_L as empirical constants. Equation 6 was derived for entangled molecules, so it should not hold for polymers whose molecular weight is less than M_e or for dilute solutions.

Lyons and Tobolsky [11] derived an equation for the viscosity of polymer solutions over the entire concentration range if the molecular weight of the polymer is less than M_e. The viscosity of the solvent, not that of the polymer, is used as the reference. Therefore, the accuracy of the theory is expected to be poor for extremely concentrated solutions [15]. However, the theory has been found to be satisfactory in a number of cases [16, 17]. The Lyons-Tobolsky equation is:

$$\frac{\eta - \eta_L}{\eta_L C[\eta]} = \exp\left[\frac{k[\eta]C}{1 - bC}\right] . \qquad (9)$$

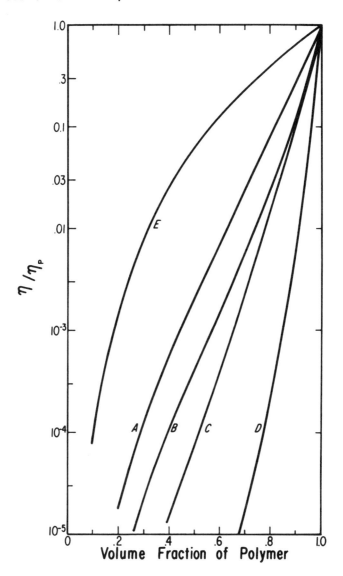

Fig. 2. Viscosity of solutions divided by the viscosity
of the polymer at the same temperature as a function of
the concentration of the polymer according to the theory of Kelley and Bueche.

	A	B	C	D	E
f_P	.1	.1	.05	.05	$f_P = f_S$
f_S	.2	.3	.10	.20	

The viscosity of the solvent is n_L, and C is the concentration of the polymer in grams/cc. The intrinsic viscosity [η] and the Huggins constant k can be determined
from the viscosity of dilute solutions of the polymer
in the solvent [18]. However, [η] and k also can be
treated as empirical constants to obtain better fit to
the experimental data. The constant b can be calculated
from melt data on the pure polymer, or it also can be
considered an empirical constant. Generally good solvents give a larger [η] and a higher dilute solution
viscosity than poor solvents do. However, in some cases
at very high polymer concentrations, poor solvents give
a higher viscosity than good solvents do [15]. The
reason for this reversal apparently is not well understood.

C. Temperature and Shear Rate Dependence of the Viscosity of Polymer Solutions

The data in Figure 1 show that the viscosity of a
polymer changes more with temperature than does the
viscosity of a liquid. The temperature dependence of
polymer solutions lies between that of the polymer and
the solvent. Thus, not only does the absolute value of
the viscosity decrease when a liquid is added to a
polymer, but the energy of activation for flow also
decreases [19].

Typical viscosity-shear rate curves for polymer
solutions of different concentrations are shown in
Figure 3 [20]. These curves show that dilute solutions
remain Newtonian in behavior to higher rates of shear
than do more concentrated solutions. The solvent reduces the number of entanglements by increasing M_e.
Reducing the number of entanglements at a given rate
of shear reduces the amount of orientation of molecular

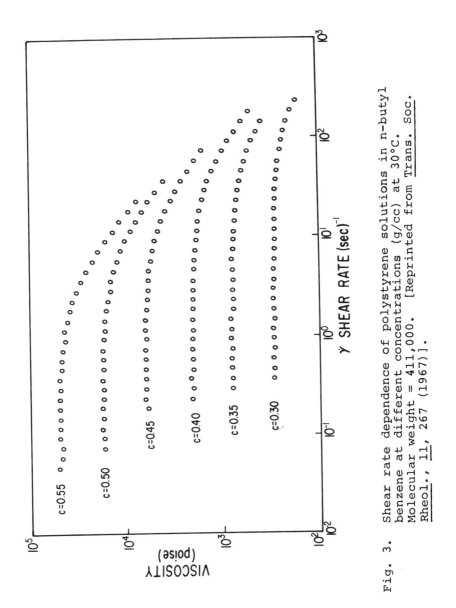

Fig. 3.   Shear rate dependence of polystyrene solutions in n-butyl
benzene at different concentrations (g/cc) at 30°C.
Molecular weight = 411,000. [Reprinted from Trans. Soc.
Rheol., 11, 267 (1967)].

segments. Since the orientation of molecular segments
is the major cause of non-Newtonian behavior, adding a
liquid to a polymer should increase the rate of shear at
which non-Newtonian behavior becomes noticeable. Behavior similar to that shown in Figure 3 has been observed
for a number of systems [17, 21-22]. Curves such as
those shown in Figures 3 to 5 can generally be shifted
vertically and horizontally to give master curves.

   D. Dynamic Rheological Properties of Solutions

   Typical dynamic rheological behavior of polymer
solutions at several concentrations are shown in Figures
4 and 5 [23]. Both the dynamic viscosity η' and the
shear modulus G' increase with concentration of polymer.
The lower the concentration the higher the frequency at
which non-Newtonian behavior, as revealed by a decrease
in viscosity with frequency, becomes apparent. At the
lowest concentrations, few if any molecular entanglements are to be expected, but the solutions still show
some elasticity and a low elastic modulus. The elasticity results from distortion of the coiled molecules by
the shearing field with resultant orientation of some of
the molecular segments. When the flow stops, the deformed molecules snap back to their normal randomly
coiled state. In concentrated solutions, entanglements
can exist which greatly enhance the elasticity. The
shear modulus increases much more rapidly than the first
power of the concentration. For example, at a frequency
of 1.0 radian/sec., a 5% solution has a modulus of less
than 10 dynes/cm^2. At the same frequency, a 10% solution has a modulus of about 400 dynes/cm^2, and a 20%
solution has a modulus of about 10,000 dynes/cm^2. Thus,
increasing the concentration by a factor of 4 increases
the shear modulus by a factor of 1000. Comparable

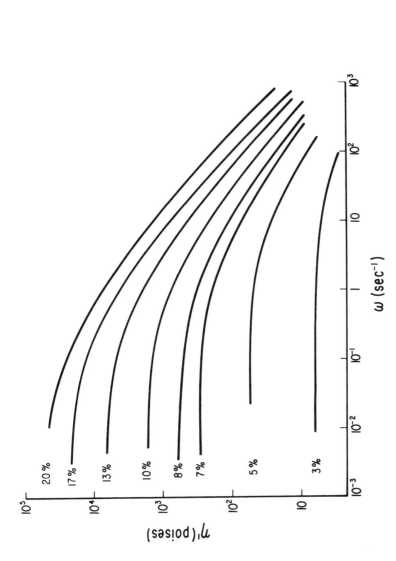

Fig. 4.   Dynamic viscosities of polyisobutylene solutions in decalin as a function of concentration and frequency. $\overline{Mn}$ = 370,000; $\overline{Mw}$ = 1.06 x 10^6. [Reprinted from J. Colloid Sci., 10, 174 (1955).]

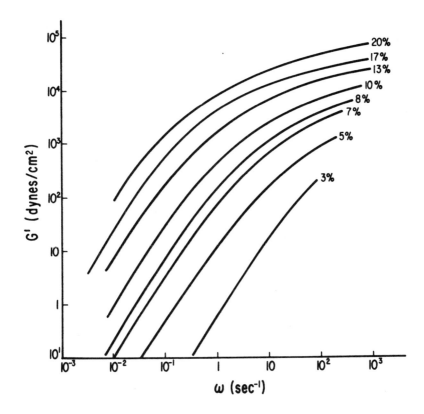

Fig. 5. Dynamic shear moduli of polyisobutylene solutions in decalin as a function of concentration
and frequency. $\overline{M}n$ = 370,000; $\overline{M}w$ = 1.06 x 10^6.
[Reprinted from J. Colloid Sci., 10, 174
(1955).]

increases in dynamic viscosity are found with increases
in the concentration at low frequencies. The changes in
both the viscosity and the modulus are much less at high
frequencies than for low frequencies when the concentration is varied.

The dynamic rheological properties have been measured for a number of polymer solutions [6, 24-27]. In
some cases the dynamic properties are reported as

dynamic compliance J' rather than G' and as loss modulus G" rather than η'.

Values of η' and G' obtained at different concentrations and temperatures can be made to give a single master curve by the use of reduced variables [5,27]. The relative viscosity η'/η_o' or the reduced modulus $G'(T_o\ C_o/TC)$ are plotted against the reduced frequency $\omega(\eta'T_o\ C_o/\eta_o'\ TC)$. The limiting dynamic viscosity at low frequencies is η_o' , T_o is the reference temperature in degrees Kelvin, while C and C_o are functions of concentration of polymer. At low concentrations, C and C_o are the concentrations of a solution and a reference solution to the first power. At higher concentrations C and C_o are equal to the square or some higher power of the concentrations of the polymer in the solutions.

II.   LUBRICANTS

Lubricants may be defined as materials which are added to polymers in small amounts to improve their processibility. The function of lubricants is to reduce the apparent viscosity of a polymer or to improve the surface appearance of a finished object made by injection molding or extrusion. There are two classes of lubricants, those which are soluble in the polymer and those which are insoluble. The soluble lubricants work primarily by reducing the viscosity and the elasticity of the polymer melt. Such materials were discussed in the first section of this chapter.

The second class of lubricants are largely insoluble in the polymer. Typical insoluble lubricants include waxes, mineral oil, metal stearates, and silicone oils. The insoluble lubricants function in a manner similar to that of a traditional lubricant in reducing mechanical friction [28-36]. Some lubricants

appear to be soluble at processing temperatures and
separate out at lower temperatures [33]. In nearly all
cases, the effect of a lubricant is shown in Figure 6.

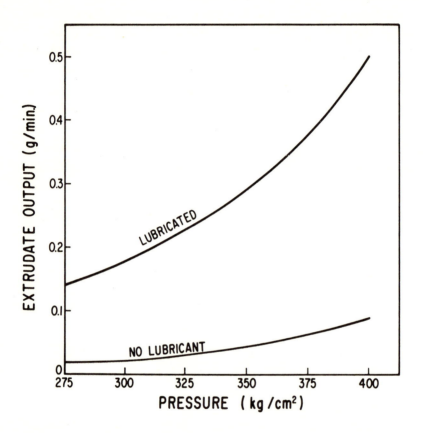

Fig. 6. Typical data on the amount of polymer
        extruded from a capillary as a function of
        the extrusion pressure for lubricated and
        unlubricated polymers.

In an extrusion process, a lubricant increases the extrusion rate at a given pressure on the melt in the extruder or injection molding machine. The effect of a lubricant in some cases is very dramatic [28].

Lubricants can change the mechanism of flow by allowing slippage to occur at a metal-polymer interface. The amount of shear within the polymer may decrease while extensional deformation increases [35]. The flow tends to be more "plug-like" in the presence of a lubricant. Unfortunately, lubricants may create some problems as well as improving the flow rate. Lubricated polymers tend to give extrudates with more die swell than the same polymers without the lubricant [35]. Lubricants also may cause a rough surface on the extrudate or cause melt fracture at lower than expected pressures. One cause of a rough surface is the sticking and then slippage of the polymer at the metal surface. The melt fracture of lubricated polymers is enhanced by the increased extensional flow that occurs in such materials.

III. REFERENCES

1. L. E. Nielsen, Mechanical Properties of Polymers and Composites, Vol. I, Marcel Dekker, New York, 1974.

2. F. Bueche, Physical Properties of Polymers, Inter-Science, New York, 1962.

3. M. Gordon and J. S. Taylor, J. Appl. Chem. (London) 2, 493 (1952).

4. M. L. Williams, R. F. Landel, and J. D. Ferry, J. Amer. Chem. Soc., 77, 3701 (1955).

5. J. D. Ferry, Viscoelastic Properties of Polymers, 2nd Ed., John Wiley, New York, 1970, p. 534.

6. T. E. Newlin, S. E. Lovell, P. R. Saunders, and J. D. Ferry, J. Colloid Sci., 17, 10 (1962).

102   LAWRENCE E. NIELSEN

7. H. Fujita and E. Maekawa, J. Phys. Chem., 66, 1053 (1962).

8. F. N. Kelley and F. Bueche, J. Polymer Sci., 50, 549 (1961).

9. H. Fujita and A. Kishimoto, J. Chem. Phys., 34, 393 (1961).

10. G. Kraus and J. T. Gruver, Trans. Soc. Rheol., 9, #2, 17 (1965).

11. P. F. Lyons and A. V. Tobolsky, Polymer Eng. Sci., 10, 1 (1970).

12. G. Pezzin, J. Appl. Polymer Sci., 10, 21 (1966).

13. G. Pezzin and N. Gligo, J. Appl. Polymer Sci., 10, 1637 (1966).

14. M. H. Cohen and D. Turnbull, J. Chem. Phys., 31, 1164 (1959).

15. K. S. Gandhi and M. C. Williams, J. Polymer Sci., C35, 211 (1971).

16. F. Rodriguez, J. Polymer Sci., B10, 455 (1972).

17. N. F. Brockmeier and S. P. Westphol, Polymer Eng. Sci., 14, 782 (1974).

18. F. Rodriguez, Principles of Polymer Systems, McGraw-Hill, New York, 1970, pp. 154-158.

19. N. Hirai, J. Polymer Sci., 40, 255 (1959).

20. W. W. Graessley, R. L. Hazleton, and L. R. Lindeman, Trans. Soc. Rheol., 11, 267 (1967).

21. L. A. Utracki, Polymer Eng. Sci., 14, 308 (1974).

22. W. W. Graessley, T. Masuda, J. E. L. Roovers, and N. Hadjichristidis, Macromol., 9, 127 (1976).

23. T. W. DeWitt, H. Markovitz, F. J. Padden, and L. J. Zapas, J. Colloid Sci., 10, 174 (1955).

24. R. F. Landel and J. D. Ferry, J. Phys. Chem., 59, 658 (1955).

25. D. M. Stern, J. W. Berge, S. F. Kurath,
     C. Sakoonkim, and J. D. Ferry, J. Colloid Sci., 17
     409 (1962).

26. N. Kinjo and T. Nakagawa, Polymer J., 5, 316 (1973).

27. J. M. Simmons, Rheol. Acta, 7, 184 (1968).

28. G. Illmann, SPE. J., 23, #6, 71 (June 1967).

29. C. L. Sieglaff, Polymer Eng. Sci., 9, 81 (1969).

30. P. L. Shah, SPE. J., 27, #1, 49 (Jan. 1971).

31. L. F. King and F. Noel, Polymer Eng. Sci., 12, 112
     (1972).

32. J. E. Hartitz, Polymer Eng. Sci., 14, 392 (1974).

33. L. L. Blyler, Jr., Polymer Eng. Sci., 14, 806 (1974).

34. P. L. Shah, Polymer Eng. Sci., 14, 773 (1974).

35. M. T. Shaw, J. Appl. Polymer Sci., 19, 2811 (1975).

36. J. L. S. Wales, J. Polymer Sci. (Sympos.), 50,
     469 (1975).

Chapter 7

NORMAL STRESSES AND DIE SWELL

I.   NORMAL STRESSES   105
II. DIE SWELL   111
III. REFERENCES   116

I. NORMAL STRESSES

Normal stresses were defined in Chapter 1. They are not found with Newtonian liquids but are character-istic of non-Newtonian polymer melts and liquids. Nor-mal stresses are primarily manifestations of the elas-ticity of polymeric materials [1-4]. It will be recalled that elasticity results from the orientation of chain segments and that chain entanglements greatly enhance the ease of orienting the segments during deformation.

The first normal stress difference, $\sigma_{11} - \sigma_{22}$, is defined as being positive if the force tends to push the retaining plates apart during shear of a fluid between plates moving parallel to one another. The normal stress differences increase with the square of the rate of shear:

$$\sigma_{11} - \sigma_{22} = \psi_1 \dot{\gamma}^2 \qquad (1)$$

$$\sigma_{22} - \sigma_{33} = \psi_2 \dot{\gamma}^2 \qquad (2)$$

The first and second normal stress coefficients, ψ_1 and ψ_2, may be functions of the rate of shear.

Normal stresses produce a number of phenomena not found with Newtonian liquids. For example, when a

105

polymer is extruded from an orifice, a capillary, or a
slit, the diameter or thickness of the resulting strand
is considerably greater than the diameter of the hole
from which it came. This phenomenon is called die swell.
Another unexpected phenomenon due to normal stresses is
the ascending of polymer liquids up rotating shafts as
illustrated in Figure 1. In contrast, rotating shafts
in Newtonian liquids cause a depression of the liquid
surface because of centrifugal forces. In cone and
plate rheometers, and in other rotating systems of similar geometry, the normal stresses tend to force the
cone and plate apart. If holes are drilled in the plates
parallel to the rotating axis, liquid will be forced up
through the holes. This is the basis of one technique
for calculating normal stresses by measuring the height
to which a liquid will ascend up a tube for a given speed
of rotation [1,5-10]. This first normal stress differ-
ence $(\sigma_{11}-\sigma_{22})$ is proportional to the height of the liquid in the tube. The liquid level is a maximum for a
tube at the center of rotation and decreases towards the
edge of the rotating disk. The normal stresses are proportional to the square of the shearing stress, so a
doubling of the speed of rotation will increase the normal stress by a factor of nearly four.

    In addition to the technique described above, nor-
mal stresses may be measured by several other methods.
When a liquid is sheared between a cone and plate or between two rotating plates, the normal stress which tends
to force the plates apart can be measured directly by a
transducer as indicated in Chapter 2 [1,5,7,8,11,12].
Normal stresses also may be estimated from the end correction used with capillary rheometers or from die swell,
or from recoverable shear [13-17]. Still another technique involves the measurement of the pressure exerted

NORMAL STRESS

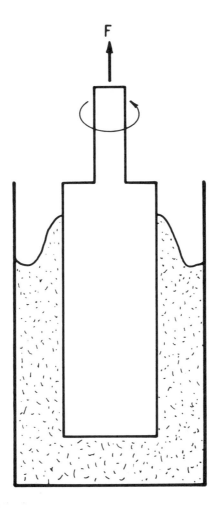

Figure 1. Creep of a polymeric liquid up the rotating
inner cylinder of a coaxial cylinder rheometer.

on the walls of a capillary in the direction perpendicular to the direction of flow [18-20].

  Normal stress measurements are generally difficult
and may be of low accuracy. The equations used to make
the calculations are complex and will not be discussed
in detail, but the equations for capillary rheometers
will illustrate the point [21,22]:

$$\sigma_{11} - \sigma_{22} = P_e + \tau_w \, (dP_e/d\tau_w) \qquad (3)$$

$$\sigma_{22} - \sigma_{33} = -\tau_w \, (dP_e/d\tau_w) \qquad (4)$$

P_e is the pressure drop at the exit of the capillary.
It is less than the P_0 of the Bagley correction (see
Chapter 2) since the Bagley correction contains both an
entrance pressure drop and an exit pressure drop.

  The first normal stress is generally positive, and
at high rates of shear may exceed the value of the shear
stress [14,23]. A positive first normal stress differ-
ence ($\sigma_{11} - \sigma_{22}$) means that the tension resulting from the
molecular orientation is parallel to the flow stream-
lines [24]. The second normal stress difference, ($\sigma_{22} -
\sigma_{33}$), is generally negative, and its value is very
small—usually about a tenth the value of the first normal stress [25-30]. However, some researchers believe
the second normal stress difference can be either zero
or positive [2,24,31]. Figure 2 shows the usual relationships for polymer melts. The literature on normal
stresses contains many inaccuracies and errors because
of incorrect measuring techniques and because of the
small magnitude of the effects in many cases. However,
most experimental studies such as those shown in Figure
3 verify the general behavior illustrated in Figure 2
[21,32]. (Note that a linear scale is used for the

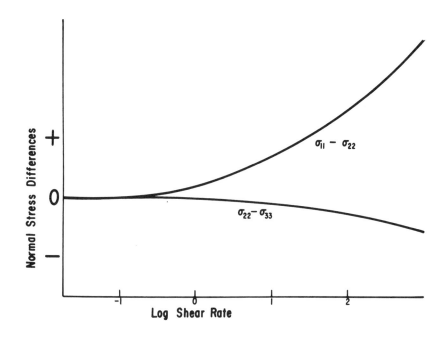

Figure 2.   General behavior of the first and second
            normal stress differences in polymer melts
            and solutions as a function of rate of shear.

normal stress difference in Figure 2 while a logarithmic
scale is used in Figure 3.)   The shear modulus shows the
same general behavior as the first normal stress difference.

    Two polymers of nearly identical viscosity charac-
teristics may have quite different behavior under actual
processing conditions.   Meissner [33] found that three
polyethylenes of nearly the same viscosities and molecular weight distributions behaved quite differently when
used to make blown film.   The rheological differences
between the polymers showed up in the first normal stress

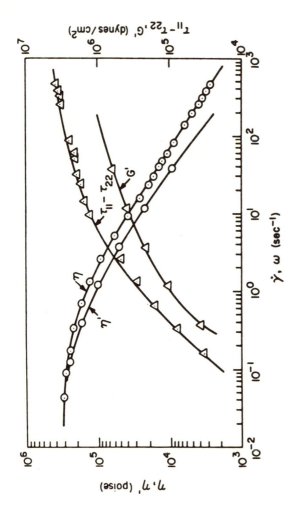

Figure 3.   Apparent viscosity η, dynamic viscosity η',
             dynamic shear modulus G', and first normal
             stress difference $\tau_{11} - \tau_{22}$ as a function of
             rate of shear $\dot{\gamma}$ or angular frequency ω for
             polystyrene at 200°C. [Reprinted from Han,
             Rheology in Polymer Processing, Academic Press,
             New York, 1976, p. 52.]

differences and in the extensional viscosities. Another
practical application where normal stresses appear to be
important is in the process of coating a wire with a
plastic. Normal stresses help produce a smooth coating
of uniform thickness up to the point where melt fracture
takes over, and at the same time they keep the wire properly centered if the second normal stress function is
negative [27]. Bird and coworkers have developed a procedure for calculating both the first normal stress difference and die swell of a polymer from the shear rate
dependence of the viscosity [34].

II.   DIE SWELL

Figure 4 illustrates how the elasticity of a liquid
produces die swell in the process of extruding a strand
through a capillary. The elastic fluid is stretched by
the shear field in the capillary and by the extensional
flow field in the entrance region to the capillary. On
leaving the capillary, the restraining forces are remov-
ed from the polymer so that the elastic material snaps
back like a stretched rubber band with a consequent increase in diameter.

Figure 5 shows the general behavior of polymer
melts on extrusion through a capillary or slit as a
function of the rate of shear. Newtonian liquids and
polymers at very low rates of shear might be expected to
have a die swell of 1.0, but actually they have a die
swell of 1.1 as measured by the radius of the extruded
strand to the diameter of the hole in the capillary [35].
Additional die swell first becomes noticeable at rates
of shear where the viscosity of the fluid starts to become non-Newtonian [36]. The die swell increases with
rate of shear in a manner similar to that of the dynamic
modulus G' or the first normal stress difference. At

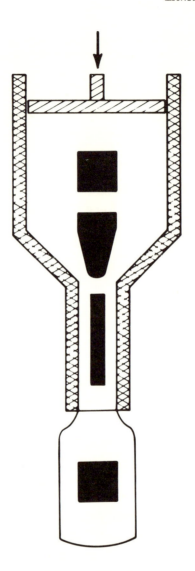

Figure 4.   Schematic diagram of how the
             tensile (elongational) forces
             act on an element of polymer in
             the entrance region to a capillary
             and how the elasticity of the melt
             causes die swell when the forces
             are removed at the end of the
             capillary.

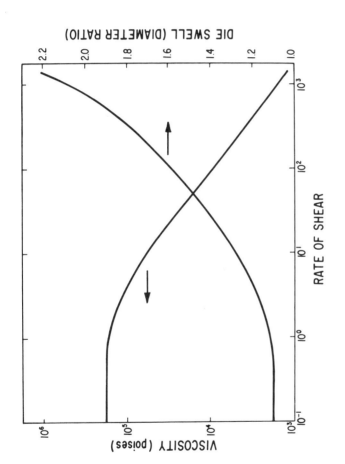

Figure 5.   General behavior of die swell and viscosity as a function of the rate of shear in a capillary or similar duct.

very high rates of shear the facts become somewhat cloudy.
In some cases, the die swell goes through a maximum at
about the shear rate at which melt fracture occurs [37,
38]. The die swell is not a unique function of the rate
of shear. For a given rate of shear, the die swell decreases as the length to diameter ratio of the capillary
increases [39-41]. The reason for this is because much
of the molecular orientation in capillary flow actually
is the result of the extensional flow in the entrance
region to the capillary and not as a result of the shear
deformation within the capillary. Thus, in a long capillary, some of the molecular orientation imparted to the
polymer in the entrance region can relax out in the capillary itself. Buchdahl, Nielsen, and Merz [39] were
the first to point out the important effect on die swell
of the time for the polymer to flow through the capillary. The shorter the time the polymer spends in the
capillary the greater is the die swell [39,41,42]. All
capillary lengths and data obtained at different temperatures give approximately a single curve when die swell
is plotted as a function of the transit time through the
capillaries. However, at a constant rate of shear, die
swell often tends to decrease somewhat as the temperature is raised because of the faster relaxation of molecular orientation at higher temperatures. However,
the maximum possible die swell as a function of shear
rate may become greater as the temperature is raised;
the maximum die swell at the higher temperature will
occur at a higher rate of shear than the maximum at a
lower temperature [38].

Factors which increase the shear modulus of a polymer melt tend to increase the die swell [43]. Thus, die
swell generally increases as the molecular weight increases when the rate of shear and temperature are held

constant [35,41]. However, there appears to be excep-
tions in which the die swell decreases as the molecular
weight increases [44]. Die swell also tends to increase
as $\overline{M}w/\overline{M}n$ increases and as long-chain branching increases [45]. This increase in die swell with increase
in long chain branching probably holds only for very
long chains with molecular weights well beyond the
critical molecular weight for entanglements.

Fillers decrease the die swell of polymers [46,47].
The effect is most pronounced with rigid fillers, but
even rubber and microgel particles, as in high impact
ABS materials, decrease the die swell [48-51]. The explanation for this effect is not entirely clear since
fillers do increase the shear modulus. Possibly the
fillers change the flow process and decrease the relative importance of extensional flow at the entrance
region to capillaries. Or possibly, fillers increase
the modulus so much that less orientation of molecular
segments results. In any case, less elastic strain
energy is stored in a polymer when filler particles are
present [49]. If the filler particles consist of agglomerates which can be broken up and dispersed on more
thorough mixing, the die swell increases with the extent
of mixing [52,53].

Various expressions have been proposed for predicting die swell [15-17,54-56]. Several of these theories
have been reviewed and evaluated [35,57]. Most of the
theories predict results which differ so little that the
experimental results do not indicate which theory is the
best. One of these theories, that of Mendelson, Finger,
and Bagley [16], relates the die swell ratio B to the
first normal stress difference $(\sigma_{11}-\sigma_{22})$ and the recoverable shear strain S_R by

$$(\sigma_{11}-\sigma_{22}) = \frac{2\tau_w}{3} (B^4 + 2B^{-2}-3)^{1/2} \qquad (5)$$

$$S_R = (B^4 + 2B^{-2} - 3)^{1/2} = \frac{2\tau_w}{3\ G} \ . \tag{6}$$

The shear stress at the wall of the capillary is τ_w and G is the shear modulus of the polymer melt. The die swell ratio B is the ratio of the radius of the extruded strand to the radius of the capillary. Other theories give similar numerical values although the form of the equations may be quite different from Equations 5 and 6.

The amount of die swell is obviously of practical importance in the spinning of fibers [58]. However, die swell is important also in other fabrication processes such as controlling the thickness of extruded sheets and in the making of bottles by blow molding. The surface roughness of extruded objects sometimes correlates with die swell; smooth surfaces are found when the die swell is small, and the rate of shear is not great enough to bring about melt fracture [59]. The anisotropy and birefringence of molded objects tend to decrease with factors which decrease the die swell [48].

III.   REFERENCES

1.   A. S. Lodge, Elastic Liquids, Academic Press, New York, 1964.

2.   K. Weissenberg, Nature, 159, 310 (1947).

3.   W. W. Graessley, Entanglement Concept In Polymer Rheology, Adv. Polymer Sci., Vol. 16, Springer-Verlag, Berlin, 1974.

4.   J. D. Ferry, Viscoelastic Properties of Polymers, 2nd Ed., John Wiley, New York, 1970.

5.   B. D. Coleman, H. Markovitz, and W. Noll, Viscometric Flows In Non-Newtonian Fluids, Springer-Verlag, New York, 1966.

6.   H. Markovitz, Proc. 4th Internat. Congr. Rheol.,
     Vol. 1, E. H. Lee and A. L. Copley, Ed., Inter-
     science, New York, 1965, p. 189.

7.   R. F. Ginn and A. B. Metzner, Proc. 4th Internat.
     Congr. Rheol., Vol. 2, E. H. Lee and A. L. Copley,
     Ed., Interscience, New York, 1965, p. 583

8.   A. Jobling and J. E. Roberts, Rheology, Vol. 2,
     F. R. Eirich, Ed., Academic Press, New York, 1958,
     p. 503.

9.   H. Markovitz and D. R. Brown, Trans. Soc. Rheol.,
     7, 137 (1963).

10. F. H. Garner, A. H. Nissan, and G. F. Wood, Trans.
     Royal Soc., A243, 37 (1950).

11. W. Philippoff and R. A. Stratton, Trans. Soc.
     Rheol., 10, 467 (1966).

12. J. D. Huppler, I. F. Macdonald, E. Ashare,
     T. W. Spriggs, R. B. Bird, and L. A. Holmes,
     Trans. Soc. Rheol., 11, 181 (1967).

13. E. B. Bagley, Trans. Soc. Rheol., 5, 355 (1961).

14. J. J. Benbow and E. R. Howells, Polymer, 2, 429
     (1961).

15. A. B. Metzner, W. T. Houghton, R. A. Sailor, and
     J. L. White, Trans. Soc. Rheol., 5, 133 (1961).

16. R. A. Mendelson, F. L. Finger, and E. B. Bagley,
     J. Polymer Sci., C35, 177 (1971).

17. Y. Mori and K. Funatsu, Appl. Polymer Symp., 20,
     209 (1973).

18. C. D. Han, M. Charles, and W. Philippoff, Trans.
     Soc. Rheol., 14, 393 (1970).

19. C. D. Han and Y. W. Kim, Trans. Soc. Rheol., 19,
     245 (1975).

20. C. D. Han, Y. W. Kim, and S. J. Chen, J. Appl.
     Polymer Sci., 19, 2831 (1975).

21. C. D. Han, Rheology In Polymer Processing,
     Academic Press, New York, 1976.

22. C. D. Han, Trans. Soc. Rheol., 18, 163 (1974).

23. W. W. Graessley, T. Masuda, J. E. L. Roovers, and N. Hadjichristidis, Macromol., 9, 127 (1976).

24. J. W. Hayes and R. I. Tanner, Proc. 4th Intern. Congr. Rheol., Vol. 3, E. H. Lee, Ed., Interscience, New York, 1963, p. 389.

25. G. Ehrmann, Rheol. Acta, 15, 8 (1976).

26. J. F. Petersen, R. Rautenbach, and P. Schümmer, Rheol. Acta, 14, 968 (1975).

27. Z. Tadmor and R. B. Bird, Polymer Eng. Sci., 14, 124 (1974).

28. R. F. Ginn and A. B. Metzner, Trans. Soc. Rheol., 13, 429 (1969).

29. O. Olabisi and M. C. Williams, Trans. Soc. Rheol., 16, 727 (1972).

30. B-L Lee and J. L. White, Trans. Soc. Rheol., 18, 467 (1974).

31. A. S. Lodge, Rheol. Acta, 4, 29 (1961).

32. C. D. Han, K. V. Kim, N. Siskovic, and C. R. Huang, Rheol. Acta, 14, 533 (1975).

33. J. Meissner, Pure Appl. Chem., 42, 553 (1975).

34. S. I. Abdel-Khalik, O. Hassager, and R. B. Bird, Polymer Eng. Sci., 14, 859 (1974).

35. J. Vlachopoulos, M. Horie, and S. Lidorikis, Trans. Soc. Rheol., 16, 669 (1972).

36. L. A. Utracki, Z. Bakerdjian, and M. R. Kamal, J. Appl. Polymer Sci., 19, 481 (1975).

37. J. A. Brydson, Flow Properties of Polymer Melts, Van Nostrand Reinhold, New York, 1970.

38. D. L. T. Beynon and B. S. Glyde, Brit. Plastics, 33, 416 (1960).

39. R. Buchdahl, L. E. Nielsen, and E. H. Merz, J. Polymer Sci., 6, 403 (1951).

40. E. B. Bagley, S. H. Storey, and D. C. West,
     J. Appl. Polymer Sci., 7, 1661 (1963).

41. M. G. Rogers, J. Appl. Polymer Sci., 14, 1679
     (1970).

42. R. L. Kruse, J. Polymer Sci., 2B, 841 (1964).

43. H. Kubota, J. Appl. Polymer Sci., 19, 2299 (1975).

44. R. A. Mendelson and F. L. Finger, J. Appl. Polymer
     Sci., 19, 1061 (1975).

45. J. E. Guillet, R. L. Combs, D. F. Slonaker,
     D. A. Weemes, and H. W. Coover, Jr., J. Appl.
     Polymer Sci., 9, 767 (1965).

46. S. Newman and Q. Trementozzi, J. Appl. Polymer Sci.,
     9, 3071 (1965).

47. N. Minagawa and J. L. White, J. Appl. Polymer Sci.,
     20, 501 (1976).

48. R. S. Hagan and D. A. Davis, J. Polymer Sci., 2B,
     909 (1964).

49. S. L. Rosen and F. Rodriguez, J. Appl. Polymer Sci.,
     9, 1601 (1965).

50. G. V. Vinogradov, V. N. Kuleznev, A. Ya. Malkin,
     A. V. Iqumnova, and O. G. Polyakov, Colloid J.
     (USSR), 29, 145 (1967) (English transl.)

51. A. Casale, A. Moroni, and C. Spreafico, ACS Adv.
     Chem. Series, No. 142, p. 172 (1975).

52. N. Tokita and I. Pliskin, Rubber Chem. Techn., 46,
     1166 (1973).

53. I. Pliskin, Rubber Chem. Techn., 46, 1218 (1973).

54. N. Nakajima and M. Shida, Trans. Soc. Rheol., 10,
     299 (1966).

55. R. I. Tanner, J. Polymer Sci., A2, 8, 2067 (1970).

56. E. B. Bagley and H. J. Duffey, Trans. Soc. Rheol.,
     14, 545 (1970).

57. V. I. Brizitsky, G. V. Vinogradov, A. I. Isaev, and
     Yu. Ya. Podolsky, J. Appl. Polymer Sci., 20,
     25 (1976).

58. I. Brazinsky, A. G. Williams, and H. L. LaNieve,
     Polymer Eng. Sci., 15, 834 (1975).

59. A. R. Berens and V. L. Folt, Polymer Eng. Sci., 9,
     27 (1969).

Chapter 8

EXTENSIONAL FLOW AND MELT FRACTURE PHENOMENA

I.   EXTENSIONAL OR ELONGATIONAL FLOW   121
II. MELT FRACTURE AND FLOW INSTABILITY   125
III. REFERENCES   129

I.   EXTENSIONAL OR ELONGATIONAL FLOW

As pointed out in Chapter 1, polymeric liquids and melts can be deformed by tensile stresses as well as by the more conventional shear stresses. The type of viscosity calculated when tensile forces are used is the tensile, elongational, or extensional viscosity, η_t.

$$\eta_t = \frac{\sigma}{\dot{\varepsilon}} = \frac{F/A}{d[\ln(L/L_0)]/dt} \qquad (1)$$

In Equation 1, σ = the tensile stress, $\dot{\varepsilon}$ = $d\varepsilon/dt$ where $\dot{\varepsilon}$ is the rate of elongation. The force at any time on a specimen which has a cross sectional area A is F. The initial length of the specimen is L_0, and the length at any time t is L.

For Newtonian liquids, the uniaxial tensile viscosity is three times the shear viscosity, while the biaxial viscosity is six times the shear viscosity [1]. However, for polymers, the tensile viscosity may be hundreds of times greater than the shear viscosity. Only in recent years has the importance of tensile viscosity been realized in spinning, polymer processing, and fabrication processes. Whenever a polymer melt or liquid is

uniaxially or biaxially stretched, extensional viscosity
becomes important. Some of these processes are spinning
of fibers, mill rolling, calendering of films, injection
molding, blow molding of bottles, and film blowing [2,3].
Uniaxial tensile deformation occurs in the spinning
process both near the entrance to the capillary or spin-
neret and in the draw down of the fiber beyond the exit
of the capillary. Extensive elongation takes place in
the entrance region to the nip of mill rolls or calender
rolls. In injection molding and extrusion machines,
elongational flow takes place whenever the cross sectional area of a channel changes , as was illustrated in
Figure 4 of Chapter 7. Biaxial extension or two-way
stretching occurs in blow molding of bottles and in the
blow extrusion of films.

Shear viscosity measurements could take the place of
the more difficult tensile measurements if the tensile
viscosity were some fixed multiplier of the shear viscosity. However, the classical paper of Ballman [4]
showed that the extensional viscosity can be hundreds of
times greater than the shear viscosity, and η_t varies in
an unpredictable manner with changes in rate of elongation or tensile stress for polymer melts.

Extensional viscosity is compared with shear viscosity in Figure 1. Three general types of tensile behavior have been reported as a function of the tensile
stress (or rate of elongation). The tensile viscosity
may be nearly independent of the tensile stress for some
polymers [4-6]. In other cases, the extensional viscosity starts to increase with tensile stress at a value
of stress about the same as where the shear viscosity
starts to decrease [5,6]. In still other cases, η_t
starts to decrease at about the same value of stress as
where the shear viscosity decreases from the Newtonian

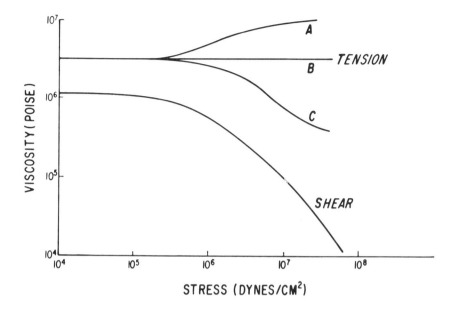

Figure 1.   Three types of tensile viscosity behaviors
compared to the shear viscosity as a function
of tensile or shear stress for polymer melts
and solutions.

value [5,7]. At present, there is no theory which is
capable of predicting this wide diversity of behavior
which is found for extensional viscosity.   Changes in
extensional viscosity are associated with non-Newtonian
behavior and the orientation of molecular segments in
the direction of elongation.   The tensile viscosity decreases with temperature, but the role of such factors
as molecular weight, entanglements, and polymer structure
in determining η_t is not clear.   Because of the lack of
a valid theory, the relationship between the uniaxial
tensile viscosity and the biaxial tensile viscosity of a
polymer also must be determined experimentally [6].

Some work indicates that the total elongation may be a more important variable than the rate of elongation [8-10].

The spinning of fibers is made easier and more stable if the elongational viscosity increases with an increase in the rate of elongation [2,3,11-14]. The reason for this is as follows: If a weak spot develops in the fiber, which results in a decrease in cross sectional area, the rate of elongation at that spot increases. The increase in rate of elongation causes an increase in the tensile viscosity which resists the stretching of the thin section.

The experimental data on the extensional viscosity of polymers and solutions have been reviewed by Dealy [1], Denson [15], and Hill and Cuculo [3]. Papers on polystyrene melts include references 4 and 9. Data on the extensional viscosity of polyethylenes can be found in references 5, 6, 13 and 16-18. Polypropylene data can be found in references 5 and 6 while data on poly-methyl methacrylate is discussed in reference 5.

Fillers affect the tensile viscosity of polymers and their solutions. Glass beads as filler in a dilute solution of polyacrylamide gave a system whose elonga-tional viscosity decreased with the rate of elongation [7]. This type of behavior also would be expected for the shear viscosity. In contrast, long fibrous fillers produce a system with very large tensile viscosities even at very low concentrations of the fibers [19]. The tensile viscosity may be hundreds of times greater than the shear viscosity at a concentration of only one volume percent. The large tensile viscosity produced by fibrous fillers has been predicted by Batchelor [20, 21]. The tensile viscosity η_t is predicted to be:

$$\eta_t = \eta \left[3 + \frac{4\phi_2 (L/D)^2}{3\ln(\pi/\phi_2)} \right] \tag{2}$$

The shear viscosity of the suspending liquid is η. The volume fraction of fibers is ϕ_2 while the length and diameter of the fibers are L and D, respectively. The tensile viscosity depends upon the square of the aspect ratio L/D. This equation, which appears to be independent of rate of elongation, assumes that the fibers are aligned in the direction of extension. In many practical situations, the fibers will not be completely oriented, but they will become more highly oriented as the rate of extension increases. In such cases, η_t should increase somewhat as $\dot{\varepsilon}$ increases. The increase in η_t according to Equation 2 is shown in Figure 2.

II. MELT FRACTURE AND FLOW INSTABILITY

At low rates of shear, polymer melts flow through capillaries, channels, and ducts to produce smooth strands. At higher rates of shear, several kinds of flow instabilities can develop in which the surface of the extruded strand becomes rough or nonuniform in cross section, and the rate of flow no longer is steady but pulsates [22-29].

The different kinds of instabilities give rise to rough or nonuniform strands from capillaries or extruders with a wide variety of appearances. "Orange peel" and "sharkskin" defects are primarily surface roughness effects. Other instabilities result in gross variations in cross sectional area. These defects include uniform pulsations in diameter, helical thread-like protusions, and very irregular jagged strands resembling the chaotic

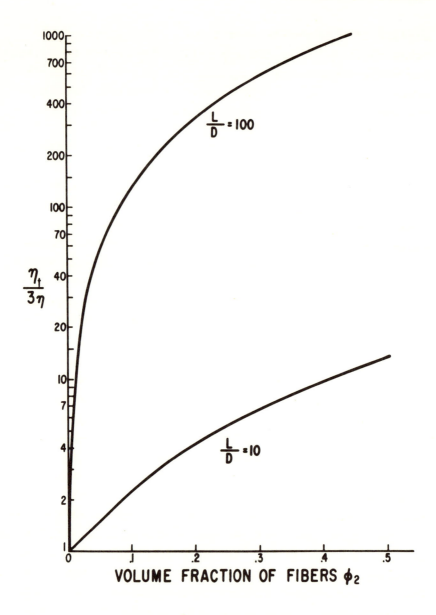

Figure 2.   The ratio of η_t to 3η for fluids containing fibers with aspect ratios of 10 and 100 as a function of the concentration of fibers. The ratio is 1.0 for Newtonian liquids.

surface of some recent lava flows. The variations in
cross sectional area are generally accompanied by pulsat-
ing variations in the extrusion pressure. Usually,
several of the above defects will be found to succeed
one another as the rate of extrusion is gradually increased.

A number of mechanisms have been proposed to explain
flow instabilities, but the exact mechanisms are still
not clear. However, all of the different kinds of strand
defects seem to be associated with melt elasticity. Two
types of phenomena may account for most of the defects.
First, the surface roughness types of defects and possibly some of the defects resulting in variations in
cross section seem to be due to a slip-stick phenomenon
at the polymer-capillary wall [30-32]. Second, the defects which result in variations in cross sectional area
are due to fracturing of the polymer melt [23,24,33-35].
Both types of phenomena may take place in some cases.
Slip-stick phenomena tend to occur in the capillary or
near the exit end of the capillary. On the other hand,
fracture phenomena generally occur in the reservoir near
the entrance to the capillary. Melt fracture generally
is the result of tensile stresses rather than shear
stresses. Tensile stresses result when the size of the
flow channel goes from a larger to a smaller cross section. This effect is illustrated in Figure 4 of Chapter
7. The change in cross section need not be abrupt but
can be a gradual taper. The less drastic the change in
cross section or the smaller the taper, the less is the
tensile elongation of polymer going through the apparatus.
If the tensile stress or the tensile elongation is too
great, the polymer melt fractures in a manner similar to
the breaking of a rubber band. When the melt fractures,
the oriented molecules cause the polymer to snap back to

an unoriented state. The orientation must then build up
before fracture can take place again. This periodic
fracture thus produces periodic variations in the appearance of the extruded strands.

The above, overly simplified, explanation enables
one to understand many of the effects observed when the
experimental conditions are changed. For example, increasing the temperature raises the onset of melt fracture to higher values of the rate of shear. Higher temperatures require higher rates of shear in order to reach
a critical fracture stress [26]. Decreasing the taper
angle at the entrance to the capillary makes it possible
to have higher shear rates in the capillary before melt
fracture occurs [25]. The effect of melt fracture becomes less noticeable as the length of the capillary is
increased because part of the effect of the fracture in
the entrance region to the capillary is damped out, and
healing can occur during the transit time through the
capillary [36]. Melt fracture occurs either at a critical shear stress or at a critical tensile stress [34,37,
38]. This fact could mean that fracture occurs at a
critical value of the shear modulus [39]. Kubota [39]
states that this critical value of G' is about 3.6×10^6
dynes/cm^2 or 3.6×10^5 Nm^{-2}. Thus, one expects the critical value of shear (or tensile) stress to decrease as the
molecular weight increases [22,28,40]. For some polymers,
at least, the critical shear stress τ_c is inversely proportional to the weight average molecular weight [22]:

$$\tau_c \doteq K/\overline{M}_w .   \eqno(3)$$

The constant K is nearly independent of temperature at
least for polystyrene. Since melt fracture relieves
part of the molecular orientation, the die swell should
decrease at stresses above the point where melt fracture
takes place. Experimentally, it has been observed that

die swell goes through a maximum and then decreases at
rates of shear near values of $\dot{\gamma}$ where melt fracture
starts [28,41].
At very high rates of shear, actual fracture of co-
valent molecular bonds can occur so that the molecular
weight of a polymer decreases. This degradation in molecular weight occurs at about the same shear stress as
that which causes melt fracture for some polymers [42].

III.   REFERENCES

1.   J. M. Dealy, Polymer Eng. Sci., 11, 433 (1971).

2.   A. B. Metzner and A. P. Metzner, Rheol. Acta, 9,
     174 (1970).

3.   J. W. Hill and J. A. Cuculo, J. Macromol. Sci. (Rev.)
     C14, 107 (1976).

4.   R. L. Ballman, Rheol. Acta, 4, 137 (1965).

5.   F. N. Cogswell, Rheol. Acta, 8, 187 (1969).

6.   C. D. Han and J. Y. Park, J. Appl. Polymer Sci., 19,
     3257 (1975).

7.   L. Nicodema, B. De Cindio, and L. Nicolais, Polymer
     Eng. Sci., 15, 679 (1975).

8.   G. Astarita, G. Marrucci, and D. Acierno, Ind. Eng.
     Chem. Fundamentals, 7, 171 (1968).

9.   A. E. Everage, Jr. and R. L. Ballman, J. Appl. Poly-
     mer Sci., 20, 1137 (1976).

10. N. E. Hudson and J. Ferguson, Trans. Soc. Rheol.,
     20, 265 (1976).

11. M. Zidan, Rheol. Acta, 8, 89 (1969).

12. A. S. Lodge, Elastic Liquids, Academic Press, New
     York, 1964.

13. H. Chang and A. S. Lodge, Rheol. Acta, 10, 448
     (1971); 11, 127 (1972).

14. Y. Ide and J. L. White, J. Appl. Polymer Sci., 20, 2511 (1976).

15. C. D. Denson, Polymer Eng. Sci., 13, 125 (1973).

16. J. Meissner, Rheol. Acta, 8, 78 (1969).

17. J. Meissner, Rheol. Acta, 10, 230 (1971).

18. M. H. Wagner, Rheol. Acta, 15, 133 and 136 (1976).

19. J. Mewis and A. B. Metzner, J. Fluid Mech., 62, 593 (1974).

20. G. K. Batchelor, J. Fluid Mech., 44, 419 (1970).

21. G. K. Batchelor, J. Fluid Mech., 46, 813 (1971).

22. R. S. Spencer and R. E. Dillon, J. Colloid Sci., 4, 241 (1949).

23. J. P. Tordella, Trans. Soc. Rheol., 1, 203 (1957).

24. J. P. Tordella, J. Appl. Phys., 27, 454 (1956).

25. E. B. Bagley and H. P. Schreiber, Trans. Soc. Rheol., 5, 341 (1961).

26. E. R. Howells and J. J. Benbow, Trans. Plast. Inst., 30, 240 (1962).

27. J. R. A. Pearson, Plastics Polymer, 37, 285 (1969).

28. J. A. Brydson, Flow Properties of Polymer Melts, Van Nostrand Reinhold, New York, 1970.

29. R. S. Lenk, Plastics Rheology, Interscience, New York, 1968.

30. J. J. Benbow, R. V. Charley, and P. Lamb, Nature, 192, 223 (1961).

31. J. J. Benbow and P. Lamb, SPE Trans., 3, 7 (1963).

32. J. M. Lupton and J. W. Regester, Polymer Eng. Sci., 5, 235 (1965).

33. E. B. Bagley and A. M. Birks, J. Appl. Phys., 31, 556 (1960).

34. A. E. Everage, Jr., and R. L. Ballman, J. Appl. Polymer Sci., 18, 933 (1974).

35. R. L. Ballman, R. L. Kruse, and W. P. Taggart,
     Polymer Eng. Sci., 10, 154 (1970).

36. J. P. Tordella, J. Appl. Polymer Sci., 7, 215 (1963).

37. L. A. Utracki, Z. Bakerdjian, and M. R. Kamal,
     Trans. Soc. Rheol., 19, 173 (1975).

38. D. R. Paul and J. H. Southern, J. Appl. Polymer Sci.,
     19, 3375 (1975).

39. H. Kubota, J. Appl. Polymer Sci., 19, 2299 (1975).

40. J. Vlachopoulos and S. Lidorikis, Polymer Eng. Sci.,
     11, 1 (1971).

41. D. L. T. Beynon and B. S. Glyde, British Plastics,
     33, 416 (1960).

42. K. B. Abbas and R. S. Porter, J. Appl. Polymer Sci.,
     20, 1289 (1976).

Chapter 9

SUSPENSIONS, LATICES, AND PLASTISOLS

I.   RIGID FILLERS - NEWTONIAN BEHAVIOR   133
II. RIGID FILLERS - NON-NEWTONIAN BEHAVIOR   142
III. RHEOLOGY OF LATICES   150
IV. RHEOLOGY OF PLASTISOLS   152
V.   REFERENCES   154

I. RIGID FILLERS - NEWTONIAN BEHAVIOR

The rheology of suspensions containing rigid fillers
is important in many areas of polymer technology. There
is a great variety of molding powders containing either
particulate or glass fiber fillers. Most rubber formula-
tions contain carbon black, and the polyvinyl chloride
in floor tiles and wire coatings contains various rigid
fillers. Latices may be suspensions of rigid polymer
particles in water, while plastisols are suspensions of
polymer particles in a liquid plasticizer. Even in the
cases where the suspending liquid is Newtonian in behavior, the presence of a filler produces profound effects
on the rheological behavior of the suspension. The
rheology becomes even more complex if the liquid phase
is non-Newtonian.

Einstein [1] first predicted the effect of a filler
on the viscosity of a Newtonian fluid. His simple equation is

$$\eta = \eta_1 (1 + k_E \phi_2) \tag{1}$$

The viscosity of the mixture is η and that of the suspending liquid is η_1. The volume fraction of filler is

ϕ_2, and k_E is the Einstein coefficient. For particles
of spherical shape, k_E is 2.5.   Table 1 shows the

Table 1

Einstein Coefficients

Type of Dispersed Phase	Orientation of Particles	Einstein Coefficient
Dispersed spheres	Any	2.50
Spherical aggregates of spheres	Any	$2.50/\phi_a$
Cubes (approximate value)	Any	3.1
Uniaxially oriented fibers	Fibers parallel to tensile stress component	$2L/D$
Uniaxially oriented fibers	Fibers perpendicular to tensile stress component	1.50

ϕ_a = maximum packing fraction of the spheres in the
aggregates

L = length of fibers with a diameter D.

Einstein coefficient for particles of various shapes and
orientations. The magnitude of the Einstein coefficient
is determined by the degree to which the particles disturb the streamlines in a flowing system. Some particle
shapes, such as rods, disturb the streamlines more than
do spheres. Although the Einstein equation is valid
only for very low concentrations of particles, it is
amazingly simple. The equation implies that the relative viscosity η/η_1 is independent of the size and
nature of the particles.

Over 100 equations have been proposed for estimating
the viscosity of a liquid containing spherical particles
up to moderate concentrations [2]. Only a few of the
better equations will be considered for predicting the
viscosity of suspensions containing particles of spherical or other shapes. One of the best and most general
equations is one proposed by Mooney [3]:

$$\ell n\,(\eta/\eta_1) = \frac{k_E\ \phi_2}{1 - \phi_2/\phi_m} \ . \qquad (2)$$

The constant ϕ_m is the maximum packing fraction. It can
be defined as

$$\phi_m = \frac{\text{True volume of filler}}{\text{Apparent volume occupied by the filler}} \ . \quad (3)$$

The maximum packing fraction can be calculated theoretically in some cases for different types of packing.
Some values of ϕ_m are listed in Table 2 [4-6]. Experimentally, ϕ_m can be estimated from sedimentation volumes
or even from the volume occupied by a given weight of a
powder. The viscosity of a suspension approaches in-
finity as the concentration approaches ϕ_m. This is to
be expected when a large number of particle-particle
contacts occur so that motion becomes restricted. Eventually, a rigid paste which has a yield point is formed
when $\phi_2 \doteq \phi_m$ [7,8]. In some cases the Mooney equation
holds well when using the theoretical or experimentally
measured values of k_E and ϕ_m for a given system. In
other cases, better fit is obtained by using empirical
values of k_E and ϕ_m. Generally, the empirical values
are close to the values expected from theory, but in
some cases the differences are large. The reasons for
this discrepancy often are not known.

Table 2

Maximum Packing Fractions ϕ_m

Particle	Type of Packing	ϕ_m
Spheres	Hexagonal close	0.7405
Spheres	Face centered cubic	0.7405
Spheres	Body centered cubic	0.60
Spheres	Simple cubic	0.524
Spheres	Random close	0.637
Spheres	Random loose	0.601
Rods	Uniaxial hexagonal close	0.907
Rods	Uniaxial simple cubic	0.785
Rods	Uniaxial random	0.82
Rods, L/D = 1	Approx. 3 dimensional random	0.704
Rods, L/D = 2	Approx. 3 dimensional random	0.671
Rods, L/D = 4*	Approx. 3 dimensional random	0.625
Rods, L/D = 8*	Approx. 3 dimensional random	0.476
Rods, L/D = 16*	Approx. 3 dimensional random	0.303
Rods, L/D = 30*	Approx. 3 dimensional random	0.173
Rods, L/D = 40*	Approx. 3 dimensional random	0.130
Rods, L/D = 50*	Approx. 3 dimensional random	0.100
Rods, L/D = 60*	Approx. 3 dimensional random	0.081
Rods, L/D = 70*	Approx. 3 dimensional random	0.065

*J. V. Milewski, Ph.D. Thesis, Rutgers University, New Brunswick, N. J., May, 1973.

An equation proposed by Dougherty and Krieger [9] has been found to predict accurately the viscosity of latices:

$$\frac{\eta}{\eta_1} = \left[1 - \phi_2/\phi_m \right]^{-k_E \phi_m} . \tag{4}$$

Nielsen [4,10] has modified a general mixture equation of the type derived by Kerner [11] for the elastic modulus of materials filled with spherical particles. This modified equation has had good success in predicting thermal and electrical conductivities, dielectric constants, and viscosities in addition to the elastic moduli of filled systems. This equation is:

$$\frac{\eta}{\eta_1} = \frac{1 + AB\phi_2}{1 - B\psi\phi_2} \tag{5}$$

where

$$A = k_E - 1 \tag{6}$$

$$B = \frac{\eta_2/\eta_1 - 1}{\eta_2/\eta_1 + 1} \doteq 1.0 \text{ for rigid fillers} \tag{7}$$

$$\psi \doteq 1 + \left(\frac{1 - \phi_m}{\phi_m^2} \right) \phi_2 . \tag{8}$$

The factor $\psi\phi_2$ is a reduced volume fraction which approaches 1.0 as ϕ_2 approaches ϕ_m. The Mooney equation, the Krieger-Dougherty equation, and the modified Kerner equation are plotted in Figure 1 for a suspension of rigid spherical particles. Random close packing ($\phi_m = 0.64$) was assumed in the calculations. All three equations approximate the Einstein equation at very low concentrations, but at concentrations above a volume

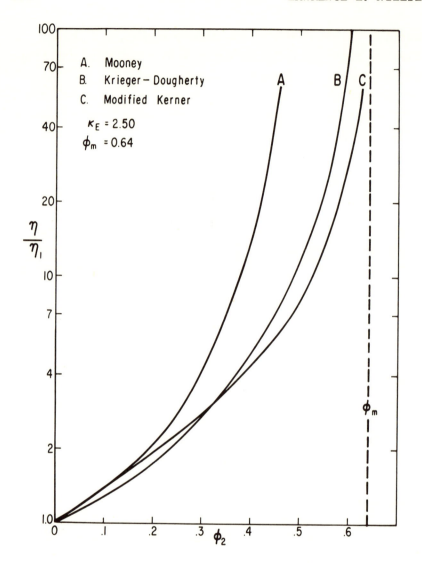

Figure 1.   Relative viscosity as a function of concentra-
            tion of spheres according to the Mooney,
            Krieger-Dougherty, and modified Kerner
            equations.

fraction of about 0.25, the Mooney equation predicts
relative viscosities considerably greater than the other
two equations. All three equations indicate that particle size has no effect on the viscosity. However, the
distribution of the size of spheres can affect the vis-
cisoty by changing ϕ_m. Mixtures containing both large
and small spheres can pack more densely than spheres of
a uniform size. As the packing density increases, ϕ_m
increases, and the relative viscosity η/η_1 decreases.

An Einstein coefficient of 2.5 is for dispersed
spheres. If the spheres form permanent, rigid aggregates, the Einstein coefficient increases because some
of the liquid becomes immobilized within the aggregates
and in the cusps where the spheres touch [12-14]. Thus,
the apparent volume of even a spherical aggregate is
larger than the true volume occupied by the sum of the
individual spheres making up the aggregate. The effect
of aggregation on the relative viscosity is illustrated
in Figure 2 for cases where the Mooney equation holds
and the type of packing is hexagonal close packing
($\phi_m = 0.74$). The Einstein coefficient for an aggregate
depends both upon the number of particles in the aggregate, as shown in Figure 3, and upon the type of packing within the aggregate [14]. In general, the Einstein
coefficient for large spherical aggregates is given by

$$k_E = \frac{2.50}{\phi_a} = \frac{2.50 \ (V_S - V_L)}{V_S} \tag{9}$$

The volume fraction of an aggregate that is actually
filled with spheres is ϕ_a while V_S is the actual volume
of the spheres making up the aggregate, and V_L is the
volume of liquid that is entrapped within and on the
surface of the aggregate. If aggregates pack in the
same manner as the individual particles do in making up

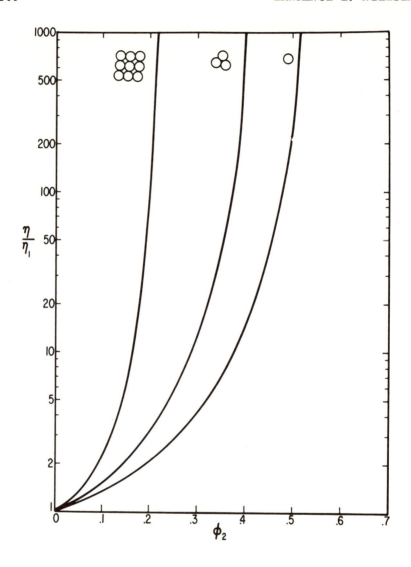

Figure 2. Effect of agglomeration of spheres on the
           relative viscosity according to the Mooney
           equation. n equals the number of particles
           in the agglomerate. [Reprinted from Nielsen,
           Mechanical Properties of Polymers and
           Composites, Vol. 2, Marcel Dekker, New York,
           1974.]

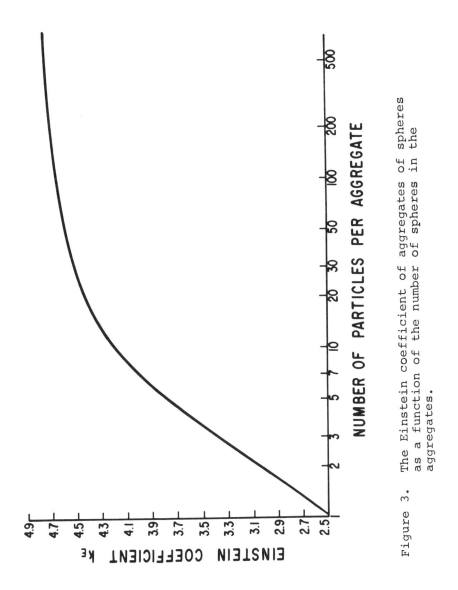

Figure 3. The Einstein coefficient of aggregates of spheres as a function of the number of spheres in the aggregates.

the aggregates,

$$\phi_m = \phi_a^2 . \tag{10}$$

For example, if spherical particles are randomly close
packed in the aggregates, $\phi_a = 0.637$. If these spherical aggregates in turn also pack in a random close manner, then the maximum packing fraction ϕ_m for this system is 0.406.

Randomly oriented rod-shaped particles and fibers
have larger Einstein coefficients than do dispersed
spheres, but the various theories differ in the values
they predict [15-19]. Part of the discrepancy arises
from the tendency of long rod-shaped particles to become
oriented in a shear field, and the importance of
Brownian motion often is unknown. Figure 4 shows the
range of Einstein coefficients for rod-like particles as
a function of their aspect ratio, i.e., their length to
diameter ratio L/D. Suspensions of rods tend to be non-
Newtonian. The viscosity decreases as the rate of shear
increases because the rods become oriented.

Oblate (flattened) spheres, flakes, and disk-shaped
particles have Einstein coefficients greater than those
of fibers of the same aspect ratio. Again, the various
theories disagree, but Figure 4 shows a typical curve of
k_E versus aspect ratio for flake-like fillers [19-21].

II. RIGID FILLERS - NON-NEWTONIAN BEHAVIOR

Even if the liquid phase is Newtonian in behavior,
filled systems become non-Newtonian and change their
properties as the rate of shear changes when the concentration of rigid filler becomes high. There are several
possible mechanisms which may be operating to bring about
changes in viscosity, give rise to true yield points, or
lead to pseudo yield points.

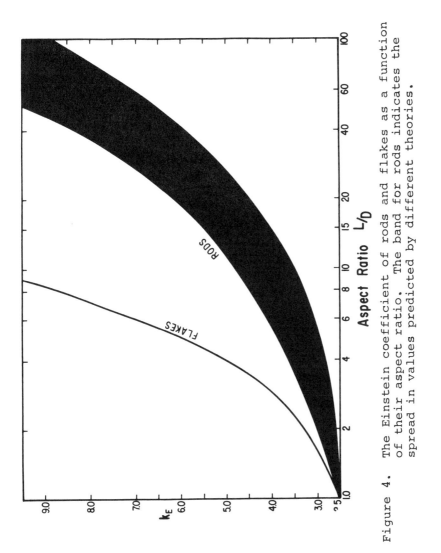

Figure 4. The Einstein coefficient of rods and flakes as a function
of their aspect ratio. The band for rods indicates the
spread in values predicted by different theories.

The break-up of weak agglomerates in a shear field
is a major cause of a decrease in viscosity as the rate
of shear increases. In a shear field, such as shown in
Figure 1 of Chapter 1 or Figure 1 of Chapter 4, a particle experiences a tensile force which tends to pull the
particle apart in a direction 45° to the plane of the
moving plates of the rheometer [22,23]. The larger the
particle the greater is the tensile force acting on it.
These agglomerates may be as small as a doublet formed
by the collision of two particles as a result of their
random Brownian motion. However, the shear rate dependence of the viscosity can not be great in this case.
The limiting viscosity for doublets at low shear rates
should not be more than about twice the limiting viscosity at very high rates of shear where only dispersed
primary particles exist. Krieger and Dougherty [24]
have developed a theory for the viscosity of rotating
collision doublets. Their equation is:

$$\eta \doteq \eta_\infty + \frac{\eta_0 - \eta_\infty}{(1 + R^3 \tau/kT)} \quad .$$ (11)

The viscosities at very low and at very high rates of
shear are η_0 and η_∞, respectively. The radius of the
primary particles is R, k is Boltzmann's constant, T is
the absolute temperature, and τ is the shear stress.
The shear stress required to break up the doublets decreases as the radius of the particles increases. This
equation has been found to hold for latices.

Most agglomerates of practical importance are much
larger than doublets. These agglomerates can be held
together by various mechanisms including van der Waals'
forces, electrostatic forces, a variety of "glues," and
by water bridges. No satisfactory theory exists for the

shear rate dependence of the viscosity caused by the
break-up of large agglomerates, although several attempts
have been made [7,25-27]. Elongational or tensile flow
fields are more effective in breaking up weak agglomerates than shear flow fields are [28].

Many concentrated suspensions are either pseudoplastic or plastic with a yield point even though the
matrix fluid is Newtonian in nature. Figure 5 illustrates plastic and pseudoplastic behavior. A plastic

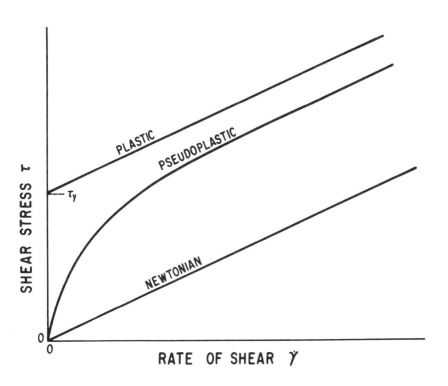

Figure 5. Typical flow curves for Newtonian, pseudoplastic, and plastic materials.

solid has a definite yield stress τ_y which must be applied before any deformation can take place. A pseudoplastic material has no yield stress, but the apparent
viscosity is very high at low rates of shear and gradu-
ally decreases to a limiting apparent consistency, $d\tau/d\dot{\gamma}$.
At high rates of shear, both plastic and pseudoplastic
materials often have a constant consistency and appear
to be Newtonian in that the shear stress is a linear
function of the shear rate.

The empirical Cross equation describes many pseudoplastic suspensions in which the apparent viscosity η_a
decreases as the rate of shear increases [27,29,30].

$$\eta_a = \eta_\infty + \frac{\eta_0 - \eta_\infty}{1 + \Omega\dot{\gamma}^m} . \qquad (12)$$

The apparent viscosities at very low and high rates of
shear are η_0 and η_∞ , respectively. The empirical constants Ω and m depend upon the system, and they are some
function of the size and the strength of the agglomerates
and how they break up in the shear field. Typical values of m are 1/2 and 2/3.

Some concentrated suspensions show a yield point.
This plastic behavior may result from some substance
which "glues" the particles of an agglomerate together
until the glue bond is broken. Plastic behavior also
can result from the strong interlocking of irregularly
shaped particles. Traces of water can collect at the
interface between particles and hold them together by
interfacial forces [7,31-34]. These water bridges between particles are probably much more important than is
generally recognized. The effect of water bridging between glass spheres in a nonpolar low molecular weight
polybutadiene polymer is illustrated in Figure 6 [7].

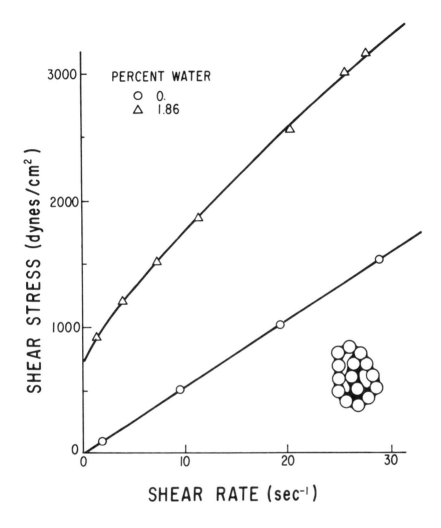

Figure 6.   The effect of water on the flow curves
of glass beads in a hydrocarbon liquid.
The insert shows how the water might
collect on the surface of the beads and
on the inside of the agglomerates to give
mechanical strength to the agglomerates.
$\phi_2 = 0.20$.

Considerable shear force is required to tear the agglomerates apart, thereby destroying the water bridges. After the bridges are broken, the suspension is much more
fluid, and, therefore, easier to shear or stir. Surfactants make flow of such systems easier by reducing
the interfacial tension, and they may change the flow to
Newtonian.

The Casson equation often holds for concentrated
plastic pastes and suspensions [35-37]. The equation is:

$$\tau^{1/2} = \tau_y^{1/2} + K\dot{\gamma}^{1/2} \ .\tag{13}$$

The yield stress is τ_y, and K is an empirical constant.
The yield stress may depend upon the amount of water
making up water bridges in agglomerates [7]. The yield
stress in other cases depends mostly upon the nature of
the particles and their concentration but not upon the
viscosity of the liquid. Generally, yield points are
found only at concentrations approaching the maximum
packing fraction of the particles in the liquid. Yield
points tend to be most pronounced in highly agglomerated
systems in which the liquid does not wet the particles
[7]. If the suspending liquid is non-Newtonian in behavior, by itself, the constant K in the Casson equation
should be modified to [37]:

$$K = K\left(\eta_a/\eta_o\right)^{1/2}\tag{14}$$

where η_o is the viscosity of the liquid at zero rate of
shear, and η_a is the apparent viscosity of the liquid at
a rate of shear $\dot{\gamma}$.

High polymer melts containing fillers are strongly
non-Newtonian in behavior because the polymer itself is

non-Newtonian in addition to the effects resulting from
the filler as discussed in this chapter. In a general
manner, the observed effects are illustrated in Figure 7.

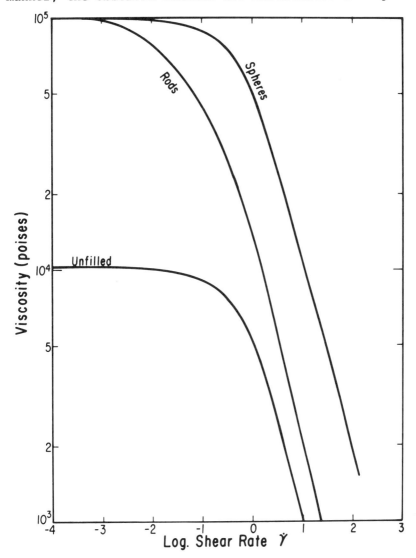

Figure 7. The effect of spherical and rod-shaped filler
particles on the flow of a polymer melt. The
concentration of spheres is higher than that
of the rods.

The fillers increase the viscosity of the polymer melt.
The shear rate dependence may be modified also, especially if the filler is fibrous or flake-like. Particles
which are not spherical tend to become oriented in a
flow field. This orientation introduces more non-
Newtonian behavior and at a lower rate of shear than for
the unfilled melt. Fillers change the elastic behavior
of polymer melts, which is most noticeable as a reduc-
tion in die swell.

III.   RHEOLOGY OF LATICES

Latices are suspensions of polymer spheres in an
aqueous medium. Latices are used directly or indirectly
in nearly all major fields and applications of polymers.
Just two of the major applications involve latex paints
and the latices used to form the rubber phase in some
high-impact polyblends.

Much of the rheology of latices has already been
discussed in the first section of this chapter under
Rigid Fillers - Newtonian Behavior. However, there are
several additional factors about latices which must be
considered which are absent or unimportant in most other
suspensions. These factors are: 1. The very small diameter of most latex particles (0.01 to 1.0 micrometers).
2. The high concentration of surfactant present. 3. The
electric charge on the particles.

Much of the surfactant is adsorbed on the surface
of the polymer particles. This adsorbed layer makes the
total particle larger than the diameter of the pure
polymer particle. Although the surfactant layer is only
of the order of 40 Å in thickness, this layer appreciably
increases the total volume when the particle is as small
as a latex particle. Thus, to calculate the viscosity of
a latex by an equation, such as the Krieger-Dougherty

equation, the volume of the surfactant adsorbed on the particles should be added to the volume of the polymer in calculating the proper volume fraction [9,38,39].

The presence of an electrical charge causes an increase in viscosity because the electrical double layer makes the particles appear larger than they actually are, and the electrical charges bring about repulsion of particles. It is difficult for one charged particle to penetrate within the electrical double layer of another. The viscosity depends upon electrolyte concentration and upon pH since the thickness of the electrical double layer depends upon concentration. The viscosity can be larger than expected for such a suspension by as much as ten to a hundred times at low ionic concentrations [9, 39-41]. As the concentration of ions increases to a concentration of the order of 0.1 mole/liter for univalent electrolytes, the viscosity goes through a minimum. At higher concentrations the viscosity tends to increase again because of agglomeration as the point of coagulation is approached. At high rates of shear, the viscosity becomes nearly independent of ionic concentration. The dependence of viscosity on rate of shear is greatest at very low concentrations of electrolyte where the thickness of the double layer is greatest and at low shear rates where the flow field has the smallest influence on distorting the double layer. These effects are illustrated schematically in Figure 8. Latices are generally Newtonian at small concentrations of polymer. The non-Newtonian behavior and shear rate dependence of viscosity generally are noticeable at volume fractions greater than about 0.25. At volume fractions of 0.5 and above, the effects discussed above become very prominent.

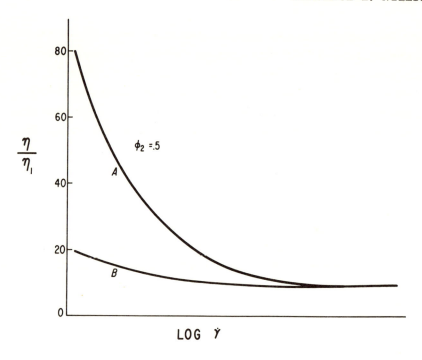

Figure 8.   Effect of ionic strength or concentration of
salts on the flow behavior of latices.
A. Very low ionic strength. B. Higher ionic
strength. $\phi_2 = 0.5$.

IV.   RHEOLOGY OF PLASTISOLS

Plastisols are suspensions of polymer particles,
such as polyvinyl chloride, in liquid plasticizers.
These suspensions, which contain relatively large amounts
of surfactants, are stable for many weeks at room temperature. At elevated temperatures, the polymer first
swells due to penetration of the plasticizer and then
dissolves and sets up into a homogeneous plasticized
elastomer. Plastisols are used to make many kinds of
objects by such techniques as rotational and slush molding. The great majority of plastisols contain polyvinyl
chloride as the polymer.

The equations for suspensions that were presented
in the first part of this chapter apply to plastisols
with some modifications [3,4,9]. Additional factors
which must be considered in the case of plastisols are:
1. Swelling of particles by the plasticizers. 2. Unusual
shear rate dependency of the viscosity and dilatancy in
which the viscosity increases with rate of shear.
3. Changes of viscosity with time of storage. 4. Unusual
temperature and time dependence of the viscosity as the
temperature is increased.

Although the layer of surfactant around the polymer
particles tends to protect the polymer from the plasticizer, the liquid gradually diffuses into the particle
and swells it. At the same time, there is less free
liquid. Therefore, the viscosity increases with time
because of both of these effects which bring about an
apparent increase in concentration of the suspended
phase [42-45]. The rate of swelling is most rapid for
plasticizers which are good solvents for the polymer.
However, the nature of the surfactant, the vapor pressure of the liquid, and its viscosity have some influence on the rate at which the viscosity of the system
increases with time. Of course, the rate of increase of
the viscosity also rapidly increases with an increase in
temperature [46]. A sudden increase in temperature will
initially lower the viscosity, of course, but the swelling of the polymer and the gelation of the system bring
about a dramatic increase in viscosity after a short
time. Eventually the plastisol sets up into a rigid
gel. At the gel point the elastic properties become
more important than the viscosity.

Some plastisols are pseudoplastic in that the apparent viscosity decreases as the rate of shear is increased [44,47]. This type of behavior has been

discussed already. Other plastisols, especially those
in which the volume fraction of polymer is greater than
0.50, are dilatant in behavior in that the apparent viscosity increases as the rate of shear is increased [44].
Several attempts have been made to develop theories of
dilatancy but with only limited success [48,49]. Dilatancy can be explained on the assumption that particle
collisions which are induced by shear bring about the
formation of agglomerates. The validity of this assumption is not known. Dilatancy can be explained also on
the assumption that other types of structure, such as
flow in layers, are developed by shear. In some plastisols, there is a discontinuity in which there is a large
jump in the apparent viscosity at a critical rate of
shear. Hoffman [50] has shown that shear produces an
ordered structure in these plastisols in which the particles form a hexagonal packing with layering parallel
to the streamlines. At the critical shear rate at which
the discontinuity in viscosity occurs, the layered hexagonal structure is destroyed. It may be that this type
of ordered structure is the cause of dilatancy in some
plastisols which do not show the discontinuity.

     In most applications for plastisols, the concentra-
tion of polymer must be very high. The viscosity might
be too great if the particles were of uniform size. Because the viscosity of most plastisols obeys a Mooney
type of equation, the viscosity can be reduced by increasing the maximum packing fraction ϕ_m by mixing particles of two or more different sizes. The small particles tend to fill the interstices between the large
particles and thus give a greater packing density.

V.   REFERENCES

1.   A. Einstein, Ann. Physik, 19, 289 (1906); 34, 591
     (1911).

2.   R. Rutgers, Rheol. Acta, 2, 305 (1962).

3.   M. Mooney, J. Colloid Sci., 6, 162 (1951).

4.   L. E. Nielsen, Mechanical Properties of Polymers and Composites, Vol. 2, Marcel Dekker, New York, 1974.

5.   J. V. Milewski, ACS Organic Coatings and Plastics Chem. Div. Preprints, 33, #2, 57 (1973).

6.   J. V. Milewski, Composites, 4, 258 (1973).

7.   S. V. Kao, L. E. Nielsen and C. T. Hill, J. Colloid Interf. Sci., 53, 358 and 367 (1975).

8.   E. C. Gay, P. A. Nelson, and W. P. Armstrong, Amer. Inst. Chem. Eng. J., 15, 815 (1969).

9.   I. M. Krieger, Adv. Colloid Interface Sci., 3, 111 (1972).

10. L. E. Nielsen, J. Appl. Phys., 41, 4626 (1970).

11. E. H. Kerner, Proc. Phys. Soc., B69, 808 (1956).

12. V. Vand, J. Phys. Colloid Chem., 52, 277 (1948).

13. T. Gillespie, J. Colloid Sci., 18, 32 (1963).

14. T. B. Lewis and L. E. Nielsen, Trans. Soc. Rheol., 12, 421 (1968).

15. J. M. Burgers, Second Report on Viscosity and Plasticity, North Holland, Amsterdam, 1938.

16. A. Okagawa, R. G. Cox, and S. G. Mason, J. Colloid Interface Sci., 47, 536 (1974).

17. H. L. Frisch and R. Simha, Rheology, Vol. 1, F. R. Eirich, Ed., Academic Press, New York, 1956, p. 525.

18. J. G. Brodnyan, Trans. Soc. Rheol., 3, 61 (1959).

19. J. Happel and H. Brenner, Low Reynolds Number Hydrodynamics, Prentice-Hall, New York, 1965.

20. G. E. Padawer and N. Beecher, Polymer Eng. Sci., 10, 185 (1970).

21. T. T. Wu, Internat. J. Solids Structures, 2, 1 (1966).

22. G. B. Jeffery, Proc. Roy. Soc., A102, 161 (1922).

23. H. L. Goldsmith and S. G. Mason, Rheology, Vol. 4,
     F. R. Eirich, Ed., Academic Press, New York, 1967,
     p. 85.

24. I. M. Krieger and T. J. Dougherty, Trans. Soc.
     Rheol., 3, 137 (1959).

25. H. D. Weymann, Proc. 4th Internat. Congr. Rheol.,
     Vol. 3, E. Lee, Ed., Wiley, New York, 1965, p. 573.

26. T. Gillespie, J. Colloid Interface Sci., 22, 563
     (1966).

27. M. M. Cross, J. Colloid Sci., 20, 417 (1965).

28. S. V. Kao and S. G. Mason, Nature, 253, 619 (1975).

29. M. M. Cross, J. Colloid Interface Sci., 33, 30 (1970).

30. M. M. Cross, Polymer Systems: Deformation and Flow,
     R. E. Wetton and R. W. Whorlow, Ed., Macmillan,
     London, 1968, p. 263.

31. H. R. Kruyt and F. G. van Selms, Recueil des
     Travaux Chim. des Pays-Bas, 62, 407, 415 (1943).

32. W. I. Higuchi and R. G. Stehle, J. Pharmaceitical
     Sci., 54, 265 (1965).

33. E. K. Fischer and C. W. Jerome, Ind. Eng. Chem., 35,
     336 (1943).

34. R. N. Weltman, Rheology, Vol. 3, F. R. Eirich, Ed.,
     Academic Press, New York, 1960, p. 189.

35. N. Casson, Rheology of Disperse Systems, C. C. Mill,
     Ed., Pergamon Press, Oxford, 1959, p. 84.

36. G. W. Scott Blair, Rheol. Acta, 5, 184 (1966).

37. T. Matsumoto, A. Takashima, T. Masuda, and S. Onogi,
     Trans. Soc. Rheol., 14, 617 (1970).

38. F. L. Saunders, J. Colloid Sci., 16, 13 (1961).

39. M. E. Woods and I. M. Krieger, J. Colloid Interface
     Sci., 34, 91 (1970).

40. J. G. Brodnyan and E. L. Kelley, J. Colloid Sci.,
     20, 7 (1965).

41. E. J. Schaller and A. E. Humphrey, J. Colloid
    Interface Sci., 22, 573 (1966).

42. D. L. Clarkson and N. D. MacLeod, Chem. Ind.,
    p. 751 (1949).

43. E. T. Severs and J. M. Austin, Ind. Eng. Chem., 46,
    2369 (1954).

44. C. Cowthra, G. P. Pearson, and W. R. Moore, Trans.
    Plast. Inst., 33, 39 (1965).

45. A. Ram and Z. Schneider, Ind. Eng. Chem. (Prod.
    Res. Dev.), 9, 286 (1970).

46. E. T. Severs and J. M. Austin, Trans. Soc. Rheol.,
    1, 202 (1957).

47. W. R. Moore and G. Hartley, Polymer Systems:
    Deformation and Flow, R. Wetton and R. Whorlow, Ed.,
    Macmillan, London, 1968, p. 285.

48. A. B. Metzner and M. Whitlock, Trans. Soc. Rheol.,
    2, 239 (1958).

49. T. Gillespie, J. Colloid Interface Sci., 22, 554
    (1966).

50. R. L. Hoffman, Trans. Soc. Rheol., 16, 155 (1972).

Chapter 10

RHEOLOGY OF POWDERS AND GRANULAR MATERIALS

I.   IMPORTANCE OF POWDER RHEOLOGY   159

II.   INSTRUMENTS   160

III. GENERAL FLOW BEHAVIOR OF POWDERS   164

IV.   POLYVINYL CHLORIDE PLASTISOL RESINS   173
      AS AN EXAMPLE OF POLYMER POWDERS

V.   REFERENCES   176

I.   IMPORTANCE OF POWDER RHEOLOGY

The flow of powders and granular materials is encountered in many aspects of everyday life. The science
of powder rheology, however, is poorly developed in
spite of its practical importance. And, compared to the
rheology of liquid-like materials, the science of the
rheology of powders and granular materials is a neglected
subject. The flow of powders is very different from the
flow of liquids. The flow characteristics of powders
often are completely different from what would be expected on the basis of experience with liquids.

The flow of powders and granular materials is important in many practical applications involving polymers. Plastic pellets and powders are stored in bins
and hoppers, and these containers should be designed to
allow the dumping of the polymers without blocking. In
recent years, powder coatings have become important and
have replaced the conventional liquid surface coatings
in many applications. The flow characteristics of such

powders are essential to the successful use of these
coatings. Polymer powders are used in certain molding
techniques such as rotational molding where flow behavior is obviously important. Granular polymers and polymer pellets are added to the first sections of extruders
and injection molding machines. The rheological properties of these granular materials before they become
melted down to a liquid may be important to the proper
performance of these machines. An appreciable fraction
of the energy used to operate extruders and injection
molding machines can go into the polymer while it is in
the granular state.

II. INSTRUMENTS

Although a few of the instruments used to study
powders will be familiar to the rheologist who works with
liquids, most of the techniques used with powders are
very different from those used with liquids. Some instruments are similar to coaxial cylinder viscometers [1-
5]. However, smooth rotor surfaces can not be used because powders can not wet the surface of the instrument
to achieve adhesion between the powder particles and the
parts of the instrument. Thus, rotor surfaces must be
roughened or contain grooves or teeth in order to trans-
mit forces into the mass of powder. Such a rotational
instrument is illustrated in Figure 1 in which the torque
required to stir the powder is measured either as a function of the speed of rotation or as a function of the
total number of turns of the rotor. Although the viscosity of a liquid is only slightly dependent upon hydrostatic pressure, the torque required to stir a powder is
extremely dependent upon the pressure exerted on the
surface of the powder. For this reason, powder rheometers
are generally equipped to apply a normal force to the

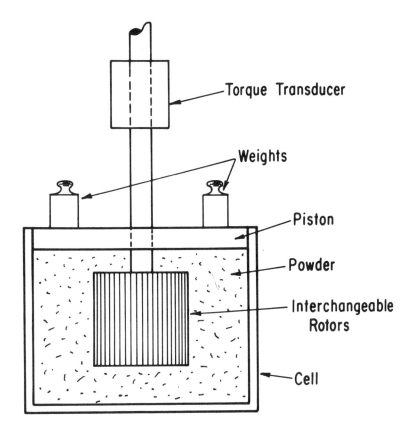

Figure 1.   Schematic diagram of a rotational rheometer
             for powders.

surface of the powder. However, this force will vary
throughout the bulk of the powder, since hydrostatic
pressure can not be transmitted uniformly through a powder as in a liquid. Rankine [1,6] derived an equation
for relating the horizontal pressure P_H resulting from
the vertical pressure P_V applied to the top surface of
a powder.

$$P_H = P_V \frac{1 - \sin \alpha}{1 + \sin \alpha} \cdot \tag{1}$$

The angle α is the internal coefficient of friction, which may sometimes be approximated by the angle of repose.

A second type of powder rheometer is a shear cell [5,7-10]. The Jenike shear cell is probably the most familiar example. One version of such an instrument is schematically shown in Figure 2. The top part of a split shear cell containing the powder is pushed at a uniform speed while the force required to produce this motion is measured by a dynamometer or force transducer. The split cell shears the powder along only a single plane. The shear force is strongly dependent upon the normal load applied to the top surface of the powder. The shear force is measured either as a function of the displacement of the top of the shear cell or as its equivalent, the time from the start of the experiment.

A third instrument measures the time for a given quantity of powder to flow through an orifice or a hollow cone with a hole at the apex [10-16]. This apparatus can be similar to a funnel with a very short stem, or it can be similar to the top half of an hour glass. Such an instrument is similar to a miniature hopper or bin. Many powders will not flow from the cone even though the size of the exit hole is many times the diameter of the particles making up the powder. Such powders form an arch very similar to the classical Roman arch which holds up the weight of the powder above the arch. It is important that hoppers and bins be designed so that arching can not stop the flow of powder from them.

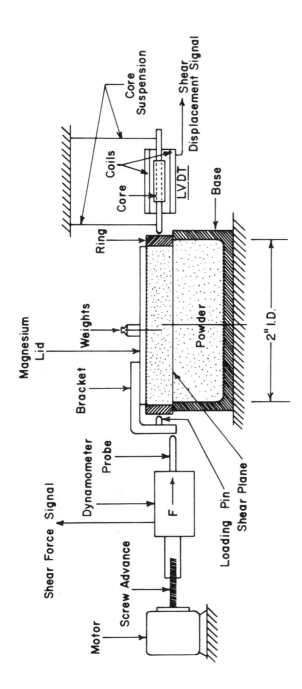

Figure 2. A modified Jenike shear cell apparatus for studying the rheology of powders
Shear cell motion is measured by the linear variable differential
transformer transducer (LVDT). [Reprinted from Takano, Nielsen, and
Buchanan, Reference 5.]

A fourth type of measurement made on powders is the angle of repose [10,17,18]. The angle of repose can be measured by many techniques. The simplest method is to form a pile of the powder on a flat surface and measure the angle formed by the sides of the pile and the horizontal surface. Intuitively, one suspects that powders which flow readily should form a smaller angle of repose than powders which do not flow so readily.

III. GENERAL FLOW BEHAVIOR OF POWDERS

The behavior of a powder in a rotational rheometer is entirely different from that of a liquid [1-5]. One of these differences is schematically shown in Figure 3. In contrast to a Newtonian liquid, the steady-state torque in a rotational rheometer generally does not increase as the rotor speed increases. For many powders, the steady-state torque remains nearly constant or may actually decrease somewhat as the speed of rotation is increased. One of the factors that determines the magnitude of the torque is the particle shape. Spherical particles tend to give a smaller torque than rough granular particles. The steady-state independence of the torque on the speed is analogous to solid-solid friction. The force required to overcome the friction of a solid block pulled across a flat solid surface is nearly independent of the speed of sliding. Thus, in some respects, the particle motion taking place in powders is more characteristic of solid-solid friction than of viscosity.

Another characteristic of powders during shearing action is the presence of a well defined slip plane or shear plane in the powder. Thus, a viscosity can not be calculated since the deformation does not vary uniformly across the annular gap of the rheometer as in the case of liquids. All the deformation generally takes place

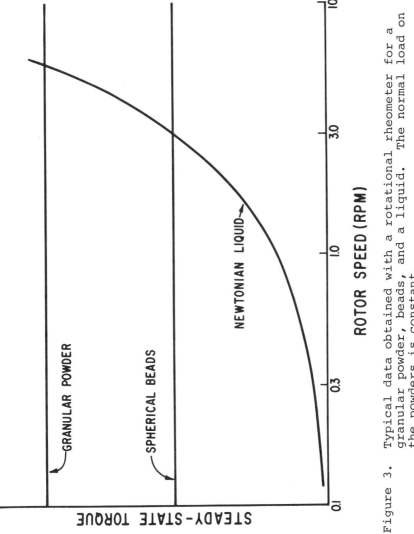

Figure 3. Typical data obtained with a rotational rheometer for a granular powder, beads, and a liquid. The normal load on the powders is constant.

within a narrow slip plane, so a true rate of shear can
not be calculated. Again, this behavior is more characteristic of solid-solid friction than of liquid viscosity.

When a rotational rheometer containing powder is
started, the steady-state torque is not achieved immedi-
ately as is illustrated in Figure 4. Several revolutions

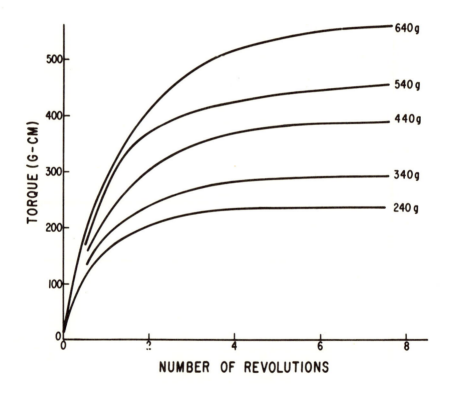

Figure 4. Rotational rheometer data with glass beads
           used as a model powder. The numbers refer
           to the normal load applied to the beads.

of the rotor are required before the torque becomes constant. Figure 4 shows still another characteristic of
powders; the torque greatly increases as the normal load
applied to the surface of the powder is increased. The
steady-state torque increases linearly with the normal
load. This behavior also is analogous to that of solid-
solid friction.

The initial density of packing of the granular
material has a strong influence on the start-up torque
of a rheometer as shown in Figure 5. Loosely packed

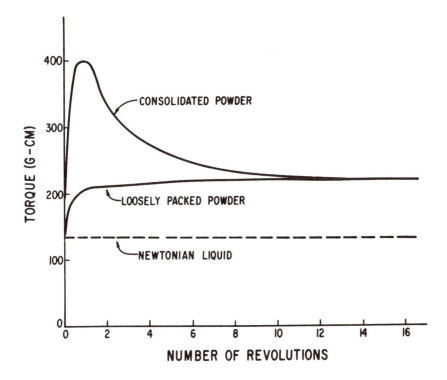

Figure 5.   Rotational rheometer data on a loosely packed
             powder and on a consolidated powder as a
             function of the number of revolutions of the
             rotor. The normal load on the powders is
             constant.

powders tend to have a monotonically increasing torque
as the number of revolutions increases. Consolidated
powders, produced by tapping or vibrating the powders to
increase the density of packing, produce a maximum in the
torque which can be considerably greater than the steady-
state torque.

Shear cells give data similar to those shown in
Figures 4 and 5, which are obtained with rotational rheometers. With shear cells, the displacement of the shear
cell takes the place of the number of rotations and the
force required to move the shear cell takes the place of
the torque in the graphs. Thus, shear cells produce data
from which shearing force is plotted against the displacement of the cell for various normal loads on the
cell.

Most powders can be classified as either cohesive
or noncohesive powders. The difference between the two
classes of powders is schematically illustrated in Figure
6. For noncohesive powders the steady-state torque in
the case of rotational rheometers or the force in the
case of a shear cell extrapolates to zero at zero normal
load on the powder. Cohesive powders have a finite yield
stress and do not extrapolate to zero steady-state torque
or displacement force as the normal load approaches zero.
However, in both cases, the torque or displacement force
is nearly a linear function of the normal load. Thus,
data from shear cells also give graphs similar to Figure
6 except the steady-state force to move the shear cell
instead of the torque is plotted against the normal load
on the powder.

Cohesive powders do not flow readily under the
action of small forces. They tend to form a uniform
coherent pile with large angles of repose. In contrast,
a noncohesive powder, such as a mass of small ball

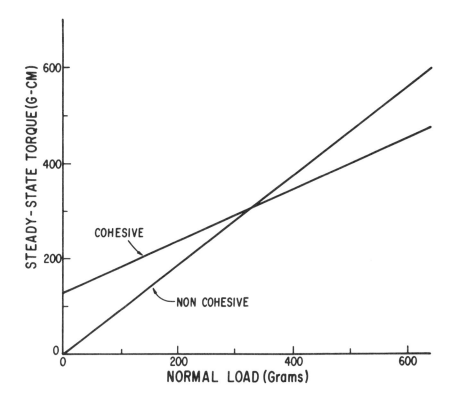

Figure 6.   Typical data obtained with a rotational
            rheometer on cohesive and noncohesive powders.
            The speed of rotation is constant.

bearings, readily flows under the action of small forces
such as gravity, and it is more difficult to form a
distinct pile of the powder.

The cause of coherence in cohesive powders can be
due to several factors. These include: 1. Rough particle
surfaces and interlocking of irregularly shaped particles.
2. Sticky coatings. 3. Interparticle attractions, such
as magnetism in iron particles or electrostatic charges
on nonconducting particles. 4. The presence of a liquid
or "glue" at points of particle-particle contacts.

Curves such as those shown in Figure 6 are examples of the Coulomb equation:

$$F_S = CA + \alpha AF_n \tag{2}$$

or

$$\tau = C + \alpha\sigma_n . \tag{3}$$

F_S is the total shear force in a shear cell, or it is the torque in a rotational rheometer. The constants C and α are the coherence of the powder and the coefficient of internal friction, respectively. The normal force or load applied to the surface of the powder is F_n. The constant A is the area of the shear cell in shear cell instruments; in rotational instruments, A is a constant dependent upon the dimensions of the rotor. The shear stress and the normal stress applied to the powder are τ and σ_n. The coefficient of internal friction is sometimes related to the angle of repose, but in general, the two quantities are not identical. In Figure 6, the intercept on the torque axis is proportional to the coherence while the slope is proportional to the coefficient of internal friction. For perfectly free-flowing powders, that is, for noncoherent powders, the coherence C is zero. Then,

$$\tau = \alpha\sigma_n . \tag{4}$$

Many cases are known, as is shown in Figure 6, where a cohesive powder has a smaller coefficient of internal friction than does a noncohesive powder. In such cases, the cohesive powder becomes easier to stir than the noncohesive powder at high normal loads. This unexpected ease of stirring cohesive powders results from their inability to pack as densely as noncohesive powders [19].

Not all powders obey the Coulomb equation, but they then follow a more general equation [9]:

$$\left(\frac{\tau}{C}\right)^n = \frac{\sigma_n}{\sigma_B} + 1.$$ (5)

The tensile strength of the powder is σ_B; it is the intercept on the horizontal axis of Figure 6 if the curve for the cohesive powder is extended in the negative normal load direction. The constant n generally varies from 1.0 to 2.0; when n = 1, the Coulomb equation is valid.

Strongly cohesive powders can not flow through orifices or from bins and hoppers. Noncohesive powders and weakly cohesive powders can flow through funnels or other types of orifices. As mentioned in the section on Instruments, short-stemmed funnels are often used to simulate crudely the flow behavior of a powder and to estimate the ease with which the powder will flow from hoppers. The rate of flow of a powder from a funnel is essentially independent of the height of the powder in the funnel [5,12]. This is in contrast to a Newtonian liquid where the rate of flow is directly proportional to the hydrostatic pressure of the liquid at the orifice.

The flow of a powder through an orifice depends upon the relative diameters of the orifice and the particles making up the powder. Figure 7 is typical of the flow behavior of powders through an orifice. There is a maximum rate of flow at some particle size. Below a critical particle size, the powder tends to form arches over the orifice which stop the flow. As the particle size approaches the diameter of the orifice, flow again becomes blocked. For a number of powders, the following equation holds approximately for the flow of powders through a circular orifice [13]:

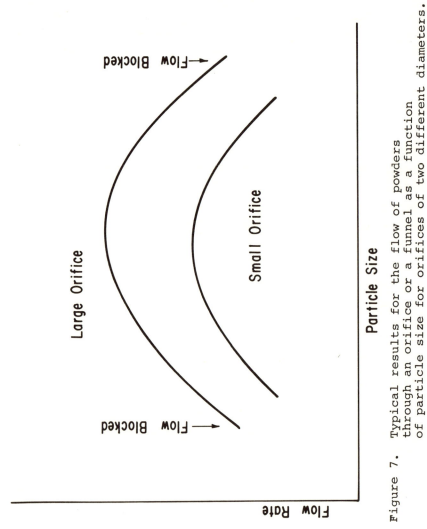

Figure 7.   Typical results for the flow of powders
through an orifice or a funnel as a function
of particle size for orifices of two different diameters.

$$D = (0.52\ D_p + 1.97) \left(\frac{4W}{60\pi\rho\sqrt{g}} \right)^{0.4} + 0.838\ D_p^{0.7} \quad (6)$$

The orifice diameter and the average particle diameter
are D and D_p, respectively. The flow rate is W(g/min),
ρ is the particle density, and g is the acceleration of
gravity. The exponent 0.4 varies somewhat for different
powders, but the exponent generally is limited between
0.3 and 0.5.

The flow behavior of powders is complex and poorly
understood. Experimental data are not very reproducible
because subtle differences in packing can produce large
differences in flow behavior. Furthermore, in most cases,
the experimental values are not related to basic characteristics of the powder, but the values depend more or
less on the type of instrument and its geometry. It
appears, however, that the flow behavior is largely controlled by defects in the packing of the powder in a manner analogous to how the deformation of crystalline solids
is controlled by dislocations and other defects such as
grain boundaries and voids [5].

IV. POLYVINYL CHLORIDE PLASTISOL RESINS AS AN EXAMPLE
OF POLYMER POWDERS

Four different polyvinyl chloride plastisol resins
will be used to illustrate how factors such as particle
shape and size can affect the flow behavior of a powdered
material [5]. Table 1 shows the bulk density and the
angle of repose of the resins. None of these fine powders,
all of which had electrostatic charges, would flow through
the ASTM standard funnel [11]. However, it is known that
polymer A flows from a large hopper more easily than
polymer B. Figure 8 shows the appearance of the particles under a microscope while Figure 9 gives the flow

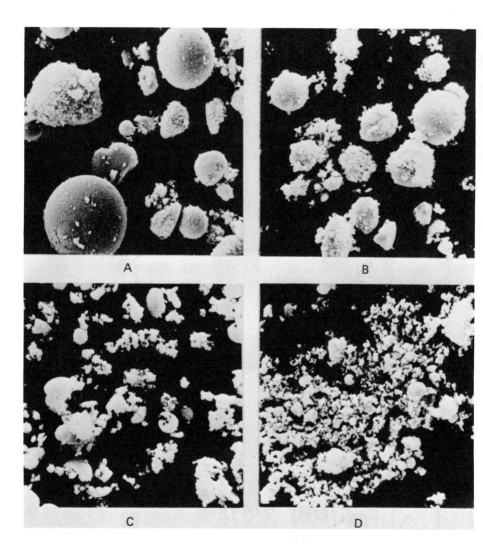

Figure 8. Particles of four PVC plastisol resins magnified 440 times. [Reprinted from Takano, Nielsen and Buchanan, Reference 5.]

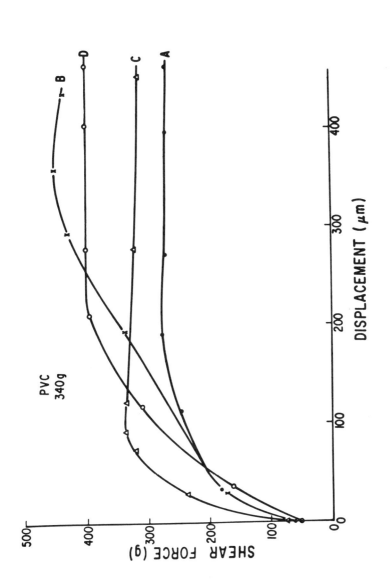

Figure 9.   Shear cell data using the four PVC plastisol resins shown in Figure 8. Normal load on the powder in a two inch diameter shear cell was 340g. [Reprinted from Takano, Nielsen, and Buchanan, Reference 5.]

Table 1

Polymer	Bulk Density (consolidated) (g/cm^3)	Angle of Repose
A	0.56	52°
B	0.43	56°
C	0.37	83°
D	0.35	72°

curves of the powders as determined in a shear cell.
Spherical particles with a relatively uniform size
(Polymer A) flow more readily than jagged particles with
a broad distribution of sizes. The flow phenomena are
so complex that it is difficult to say much more. There
is little correlation of the flow behavior with density,
angle of repose, and particle shape or size. As expected, the jagged particles do not pack as well as the
smoother ones. Less expected is the trend for the angle
of repose to increase as the density decreases.

V.   REFERENCES

1. M. M. Benarie, Brit. J. Appl. Phys., 12, 514 (1961).

2. H. Kuno and K. Kurihara, Rheol. Acta, 4, 73 (1965).

3. H. Kuno and M. Senna, Rheol. Acta, 6, 284 (1967).

4. R. Rautenbach and E. Goldacker, Kunststoffe, 61,
    485 (1971).

5. M. Takano, L. E. Nielsen and R. W. Buchanan,
    Organic Coatings and Plastics Chem. Div., ACS,
    Preprints, 33, #2, 447 (1973).

6. W. J. M. Rankine, Phil. Trans. Roy. Soc. London, 146,
    9 (1856).

7. A. W. Jenike, Gravity Flow of Bulk Solids, Bull.
    No. 108, Utah Engineering Station, University of
    Utah, Salt Lake City, 1961.

8. A. W. Jenike, P. J. Elsey, and R. H. Wooley,
    Proc ASTM, 60, 1168 (1960).

9. M. D. Ashton, D. C-H.Cheng, R. Farley, and
    F. H. H. Valentin, Rheol. Acta, 4, 206 (1965).

10. R. L. Brown and J. C. Richards, Principles of Powder
    Mechanics, Pergamon Press, Oxford, 1970.

11. ASTM Test D-1895-61T, ASTM Stds., Pt. 27, 609
    (1964).

12. R. T. Fowler and J. R. Glastonbury, Chem. Eng. Sci.,
    10, 150 (1959).

13. T. M. Jones and N. Pilpel, J. Pharm. Pharmac., 18,
    81 (1966).

14. C. F. Harwood and N. Pilpel, Chem. Process Eng.,
    49, #7, 92 (1968).

15. M. Ahmad and N. Pilpel, Rheol. Acta, 8, 448 (1969).

16. W. Reisner, Powder Tech., 1, 257 (1968).

17. D. Train, J. Pharm. Pharmac., 19, 127T (1958).

18. R. L. Brown, Powders In Industry, Soc. Chem. Ind.
    Monograph #14, London, 1961, p. 150.

19. H. H. Hausner, Int. J. Powder Metallurgy, 3, #4,
    7 (1967).

APPENDIX

List of Symbols

A   Cross sectional area

A   A constant in modified Kerner equation (Chap. 9)

A   Area of a shear cell (Chap. 10)

a_T Shift factor (Chap. 3)

b   A constant (Chap. 6)

B   Die swell ratio = (diameter of extruded strand)/ (diameter of capillary)

B   A constant of modified Kerner equation (Chap. 9)

C   Concentration of polymer in g/cc

C   A constant (Chap. 3)

C   Coherence of a powder (Chap. 10)

D   Diameter of a sphere

D   Distance between shearing surfaces (Chap. 1)

D   Diameter of a fiber

D   Diameter of orifice

D_p Average diameter of a powder particle

e   Bagley correction factor for flow through a capillary

E   Energy of activation for flow

f_L Free volume of a liquid (Chap. 6)

f_p Free volume of a polymer (Chap. 6)

F   Force applied perpendicular to plates of a parallel plate rheometer

F   Tensile force applied to a specimen to stretch it

F_n Normal or vertical force applied to a powder

F_s Total shear force applied to a shear cell (Chap. 10)

179

g   Acceleration of gravity (Chap. 10)

G   Shear modulus

G' Real part of dynamic shear modulus

G" Loss modulus; imaginary part of the dynamic shear modulus

G* Complex or dynamic shear modulus

h   Length of inner cylinder of a coaxial cylinder rheometer

h   Separation of plates in a parallel plate rheometer

h_o Initial spacing of plates in a parallel plate rheometer

h_o A correction for the length of cylinder in a coaxial cylinder rheometer

I   An interaction term (Chap. 6)

J' Real part of dynamic compliance

k   A constant

k   Instrument calibration constant (Chap. 2)

k   Huggin's constant (Chap. 6)

k   Boltzmann's constant (Chap. 9)

k_E Einstein coefficient for viscosity of suspensions

K   A constant

K α_L/α_P (Chap. 6)

K   A constant of the power law equation (Chap. 4)

K_1 A constant (Chap. 5)

K_2 A constant (Chap. 5)

L   Length of a capillary

L   Stretched length of a tensile specimen

L   Length of a fiber

L   Major axis of an ellipsoid (Chap. 4)

L_o Unstretched length of a specimen

m   A constant (Chap. 4,9)

M   Torque

M_e Molecular weight between entanglement points

M_e^o Molecular weight between entanglement points for pure polymer

$\overline{M}_n$ Number average molecular weight

M_o Molecular weight of monomeric unit

$\overline{M}_w$ Weight average molecular weight

n   A constant of the power law equation for a fluid

N   Normal force trying to separate a cone and plate during rotation of cone

P   Pressure

P_e Pressure drop at end of a capillary

P_H Horizontal component of pressure in a powder

P_V Vertical component of pressure in a powder

Q   Volumetric flow rate through a capillary

R   Radius of a capillary, or of a cone, or of a cylinder (Chap. 2)

R   Gas constant (Chaps. 3 and 5)

R   Radius of a sphere (Chap. 9)

S   Amount of shear displacement (Chap. 1)

S_R Recoverable shear strain (Chap. 7)

t   Time

T   Temperature (°K generally)

T_g Glass transition temperature

T_{gL} Glass transition temperature of solvent

T_{gP} Glass transition temperature of polymer

T_o A reference temperature (°K)

$\overline{v}$ Average velocity of flow across a capillary diameter

V   Total volume of fluid extruded through a capillary

V   Total volume of fluid in a parallel plate rheometer

V_f Free volume of polymer (Chap. 3)

V_L Volume of liquid entrapped within and on the surface of an aggregate (Chap. 9)

V_O Close packed volume (Chap. 3)

V_s Actual volume of spheres making up an aggregate (Chap. 9)

W   Minor axis of an ellipsoid (Chap. 4)

W   Flow rate of a powder through an orifice (g/min), (Chap. 10)

X_i Contribution of group i to the energy of activation of a monomeric unit made up of several groups (Chap. 3)

α Angle in radians which cone makes with the plate in a cone-plate rheometer (Chap. 2)

α Internal coefficient of friction in powders (Chap. 10)

α_L Difference between volume coefficients of expansion above and below T_g of solvent

α_p Difference between volume coefficients of expansion above and below T_g of polymer

γ Shear strain

$\dot{\gamma}$ Rate of shear, $d\gamma/dt$

γ_O Amplitude of shear strain in a dynamic rheometer

$\dot{\gamma}_w$ Rate of shear at wall of a capillary

δ Loss angle defined by $\tan \delta = \eta'/\eta'' = G''/G'$

ε Tensile strain or elongation

$\dot{\varepsilon}$ Rate of tensile elongation

η Viscosity

$[\eta]$ Intrinsic viscosity $[\eta] = \lim\limits_{c \to 0} (\eta/\eta_L - 1)/C$

$\eta*$ Complex dynamic viscosity

η' Real part of dynamic viscosity

η'' Imaginary part of dynamic viscosity

η_1 Viscosity of suspending medium (continuous phase)

η_2 Viscosity of dispersed phase

η_O Newtonian viscosity at zero rate of shear

η_∞ Viscosity at infinite rate of shear

η_a Apparent viscosity

η_c Consistency

η_g Viscosity at glass transition temperature

η_L Viscosity of a liquid

η_p Viscosity of a polymer

η_t Elongational, extensional, or tensile vsicosity

μ Interfacial tension

ρ Density of polymer

ρ Density of polymer at T_g

σ Tensile stress

σ_B Tensile strength of a powder

σ_n Normal or vertical stress applied to a powder

$\sigma_{11} - \sigma_{22}$ First normal stress difference

$\sigma_{22} - \sigma_{33}$ Second normal stress difference

τ Shear stress

τ_c Critical shear stress for melt instability

τ_o Amplitude of shear stress in a dynamic rheometer

τ_w Shear stress at wall of a capillary

τ_y Yield stress

ϕ_1 Volume fraction of continuous phase

ϕ_2 Volume fraction of dispersed phase

ϕ_a Maximum packing fraction of particles within aggregates

ϕ_L Volume fraction of liquid or solvent

ϕ_m Maximum packing fraction of particles in a suspension

ϕ_p Volume fraction of polymer

ψ A reduced concentration factor (Chap. 9)

ψ_1 First normal stress coefficient (Chap. 7)

ψ_2 Second normal stress coefficient (Chap. 7)

ω Angular frequency in radians/second

Ω A constant (Chaps. 4 and 9)

AUTHOR INDEX

Numbers in parentheses are reference numbers and indicate that an author's work is referred to although his name is not cited in text. Underlined numbers give the page on which the complete reference is listed.

A

Abbas, K. B., 129 (42), 131

Abdel-Khalik, S. I., 111 (34), 118

Acierno, D., 124 (8), 129

Ahmad, M., 162 (15), 177

Allen, V. R., 74 (17), 84

Armstrong, W. P., 135 (8), 155

Arnold, K. R., 63 (71), 68

Ashare, E., 106 (12), 117

Ashton, M. D., 162, 171 (9), 177

Astarita, G., 124 (8), 129

Austin, J. M., 153 (43) 153 (46), 157

B

Bagley, E. B., 16 (7), 16 (8), 27, 72 (9), 84 106 (13), 106, 115 (16), 117, 114 (40), 115 (56), 119, 125, 128 (25), 127 (33), 130

Bakerdjian, Z., 111 (36), 118, 128 (37), 131

Ballman, R. L., 23 (17), 28, 40 (10), 44, 62 (57), 62 (60), 68, 72, 74 (10), 72 (11), 84, 122, 124 (4), 124 (9), 129, 127, 128 (34), 130, 127 (35), 131

Bartok, W., 56 (38), 67

Batchelor, G. K., 124 (20), 124 (21), 130

Beecher, N., 142 (20), 155

Belcher, H. V., 33 (3), 44

Benarie, M. M., 160, 161, 164 (1), 176

Benbow, J. J., 50 (14), 65, 106, 108 (14), 117, 125, 128 (26), 127 (30), 127 (31), 130

Berens, A. R., 83 (45), 83 (46), 86, 116 (59), 120

Berge, J. W., 24 (29), 28, 98 (25), 103

Bergen, Jr., R. L., 62 (56), 67

Bernhardt, E. C., (1), 9

185

Berry, G. C., 81, 82 (32), 85

Bestul, A. B., 33 (3), 44

Beynon, D. L. T., 114 (38), 118, 129 (41), 131

Bird, R. B., 106 (12), 117, 108, 111 (27), 111 (34), 118

Birks, A. M., 127 (33), 130

Birnboim, M. H., 24 (28), 28, 24 (32), 29

Bishop, E. T., 63 (69), 68

Blyler, Jr., L. L., 54 (28), 66, 82 (42), 86, 99, 100 (33), 103

Boghetich, L., 82 (36), 85

Brazinsky, I., 116 (58), 120

Brenner, H., 56 (37), 66, 59 (46), 67, 142 (19), 155

Brizitsky, V. I., 115 (57), 119

Brockmeier, N. F., 92, 96 (17), 102

Brodnyan, J. G., 142 (18), 155, 151 (40), 156

Brown, D. R., 106 (9), 117

Brown, R. L., 162, 164 (10), 164 (18), 177

Brydson, J. A., (2), 9, 48 (1), 65, 114 (37), 118, 125, 128, 129 (28), 130

Buchanan, R. W., 160, 162, 164, 171, 173 (5), 176

Buchdahl, R., 18 (9), 18 (10), 27, 49 (8), 49 (9), 49 (10), 50 (13), 65, 114 (39), 118

Bueche, F., 40 (11), 44, 49 (11), 51 (16), 51 (17), 51 (18), 65 69 (7), 84, 88, 89 (2), 101, 90 (8), 102

Burgers, J. M., 142 (15), 155

                                    C

Cancio, L. V., 82 (38), 85

Carley, J. F., 43 (23), 45

Casale, A., 40 (15), 45, 61 (52), 67, 69, 72, (4), 72 (8), 84, 115 (51), 119

Casson, N., 148 (35), 156

Chaffey, C. E., 59 (46), 67

Chang, H., 124 (13), 129

Charles, M., 108 (18), 117

Charley, R. V., 127 (30), 130

Chartoff, R. P., 24 (33), 29

Chee, K. K., 74 (16), 84

Chen, S. J., 62 (62), 68, 108 (20), 117

Cheng, C-H., 162, 171 (9), 177

Chii, N. V., 63 (74), 68

Chung, C. I., 63 (73), 68

Civordi, E., 72 (8), 84

Clarkson, D. L., 153 (42), 157

Cogswell, F. N., 23 (18), 28, 43 (22), 45, 63 (72), 68, 122, 123, 124 (5), 129

Cohen, M. H., 42 (21), 45, 90 (14), 102

Coleman, B. D., 106 (5), 116

Collins, E. A., 83 (47), 86

Colwell, R. E., (11), 10, 11, 12, 18, 22 (1), 12 (4), 27, 48 (2), 65

Combs, R. L., 115 (45), 119

Cooper, S. L., 63 (70), 68

Coover, Jr., H. W., 115 (45), 119

Cowthra, C., 153, 154 (44), 157

Cox, R. G., 142 (16), 155

Cox, W. P., 54 (29), 66, 80 (26), 85

Cross, A. H., 20 (12), 28

Cross, M. M., 51, 52 (23), 66, 145, 146 (27), 146 (29), 146 (30), 156

Cuculo, J. A., 122, 124 (3), 129

D

Davis, D. A., 61 (54), 67, 115, 116 (48), 119

Dealey, J. M., 23 (16), 28, 54 (33), 66, 121, 124 (1), 129

De Cindio, B., 123, 124 (7), 129

Denson, C. D., 23 (19), 23 (23), 28, 124 (15), 130

De Witt, T. W., 24 (26), 28, 96 (23), 102

Dienes, G. J., 22 (15), 28

Dillon, R. E., 125, 128 (22), 130

Dougherty, T. J., 143 (24), 156

Douy, A., 61 (49), 67

Duffey, H. J., 115 (56), 119

E

Economy, J., 82 (44), 86

Ehrmann, G., 108 (25), 118

Einstein, A., 56 (35), 66, 133 (1), 154

Eirich, F. R., (3), 10

Elsey, P. J., 162 (8), 177

Elyash, L. J., 24 (32) 29, 40 (16), 45

Estes, G. M., 63 (70), 68

Everage, Jr., A. E., 62 (59), 68, 124 (9), 129, 127, 128 (34), 130

F

Farley, R., 162, 171 (9), 177

Ferguson, J., 124 (10), 129

Ferry, J. D., 24 (24), 24 (28), 24 (30), 28, 33 (4), 33, 36 (5), 44, 52 (25), 66, 80 (27), 85, 89 (4), 89, 99 (5), 90, 98 (6), 101, 98 (24), 102, 98 (25), 103, 105 (4), 116

Fikhman, V. D., 23 (20), 28

Finger, F. L., 106, 115 (16), 117, 115 (44), 119

Fischer, E. K., 146 (33), 156

Fitzgerald, E. R., 24 (30), 28

Fitzgerald, W. E., 82 (43), 86

Flory, P. J., 49 (5), 65, 69 (1), 69 (2), 83, 75 (23), 85

Folt, V. L., 63 (66), 68, 83 (45), 83 (46), 86, 116 (59), 120

Fowler, R. T., 162, 171 (12), 177

Fox, T. G., 69 (1), 69 (2), 69 (3), 83, 74 (17), 84

Frazer, W. J., 40 (16), 45

Friedman, E. M., 74 (15), 84

Frisch, H. L., 56 (36), 66, 142 (17), 155

Fujii, T., 54 (30), 66

Fujita, H., 90 (7), 90 (9), 102

Funatsu, K., 106, 115 (17), 117

G

Gale, J. C., 63 (73), 68

Gallo, R. J., 23 (23), 28

Gallot, B., 61 (49), 67

Gandhi, K. S., 92, 94 (15), 102

Garner, F. H., 106 (10), 117

Gay, E. C., 135 (8), 155

Gillespie, T., 139 (13), 155, 145 (26), 156, 154 (49), 157

Ginn, R. F., 106 (7), 117, 108 (28), 118

Glastonbury, J. R., 162, 171 (12), 177

Gligo, N., 90 (13), 102

Glyde, B. S., 114 (38), 118, 129 (41), 131

Goldacker, E., 160, 164 (4), 176

Goldsmith, H. L., 56, 57, 59 (40), 67, 143 (23), 156

Gordon, M., 88 (3), 101

Graessley, W. W., 49, 51 (12), 65, 51 (19), 51 (20), 66, 72 (12), 72, 74 (13), 84, 81 (34), 85, 94 (20), 96 (22), 102, 105 (3), 116, 108 (23), 118

Gratch, S., 69 (3), 83

Gruver, J. T., 40 (12), 44, 63 (68), 68, 75 (21), 84, 81 (33), 85, 90 (10), 102

Guillet, J. E., 115 (45), 119

Guth, E., 49 (4), 65

H

Haas, T. W., 82 (42), 86

Hadjichristidis, N., 81 (34), 85, 96 (22), 102, 108 (23), 118

Hagan, R. S., 61 (54), 67, 115, 116 (48), 119

Han, C. D., (4), 10, 62 (62), 62 (63), 68, 108 (18), 108 (19), 108 (20), 108 (21), 117, 108 (22), 108 (32), 118, 122, 123, 124 (6), 129

Hanson, D. E., 63 (72), 68

Happel, J., 56 (37), 66, 142 (19), 155

Harding, S., 40 (11), 44, 51 (17), 65

Harper, R., 24 (26), 28

Hart, G. M., 82 (41), 86

Hartitz, J. E., 99 (32), 103

Hartley, G., 153 (47), 157

Harwood, C. F., 162 (14), 177

Hashimoto, T., 61 (47), 61 (48), 67

Hassager, O., 111 (34), 118

Hausner, H. H., 170 (19), 177

Hayes, J. W., 108 (24), 118

Hazleton, R. L., 51 (20), 66, 72 (12), 84, 94 (20) 102

Herrmann, H. D., 40 (13), 45

Higuchi, W. I., 146 (32), 156

Hildebrand, J. H., 42 (19), 45

Hill, C. T., 135 , 145, 146 , 148 (7), 155

Hill, J. W., 122, 124 (3), 129

Hirai, N., 94 (19), 102

Hobbs, L. M., 81, 82 (32), 85

Hoffman, R. L., 154 (50), 157

Hoftyzer, P. J., 35, 36 (6), 44

Hogan, J. P., 81, 82 (35), 85

Holden, G., 63 (69), 68

Holmes, L. A. 106 (12), 117

Horie, M., 111, 115 (35), 118

Houghton, W. T., 106, 115, (15), 117

Howells, E. R., 50 (14), 65, 106, 108 (14), 117, 125, 128
   (26), 130

Huang, C. R., 108 (32), 118

Hudson, N. E., 124 (10), 129

Humphrey, A. E., 151 (41), 157

Huppler, J. D., 106 (12), 117

Huseby, T. W., 54 (28), 66

                                   I

Ibaragi, T., 54 (31), 66, 75, 80 (22), 84

Ide, Y., 124 (14), 130

Igumnova, A. V., 61 (51), 67, 115 (50), 119

Illmann, G., 99 (28), 103

Inoue, T., 61 (47), 61 (48), 67

Isayev, A. I., 63 (74), 68, 115 (57), 119

J

Jakob, M., 63 (67), 68
James, H. M., 49 (4), 65
Jeffery, G. B., 57 (42), 67, 143 (22), 156
Jenike, A. W., 162 (7), 162 (8), 177
Jerome, C. W., 146 (33), 156
Jobling, A., 106 (8), 117
Johnson, J. F., 31 (2), 44, 40 (15), 45, 69, 72 (4), 84
Jones, T. M., 162, 171 (13), 177
Jung, A., 43 (25), 45

K

Kamal, M. R., 43 (26), 45, 111 (36), 118, 128 (37), 131
Kao, S. V., 135, 145, 146, 148 (7), 155, 145 (28), 156
Kato, H., 54 (30), 66
Kawai, H., 61 (47), 61 (48), 67
Keeney, R., 80 (26), 85
Kelley, E. L., 151 (40), 156
Kelley, F. N., 90 (8), 102
Kerner, E. H., 137 (11), 155
Kim, K. U., 108 (32), 118
Kim, K. Y., (11), 10, 11, 12, 18, 22 (1), 27, 48 (2), 65,
    62 (62), 62 (63), 68
Kim, Y. W., 108 (19), 108 (20), 117
King, L. F., 99 (31), 103
Kinjo, N., 98 (26), 103
Kirchevskaya, I. Yu., 63 (74), 68
Kishimoto, A., 90 (9), 102
Kitagawa, K., 69, 75 (6), 84, 80 (24), 85
Klemm, H. F., 22 (15), 28
Knappe, W., 40 (13), 45
Kratz, R. F., 82 (36), 85
Kraus, G., 40 (12), 44, 63 (68), 68, 75 (21), 84, 81 (33),
    85, 90 (10), 102

Krieger, I. M., 14 (6), 27, 137, 151, 153 (9), 155, 143, (24), 151 (39), 156

Krier, C. A., 83 (47), 86

Kruse, R. L., 114 (42), 119, 127 (35), 131

Kruyt, H. R., 146 (31), 156

Kubota, H., 62 (55), 67, 114 (43), 119, 128 (39), 131

Kuchinka, M. Yu., 63 (67), 68

Kuleznev, V. N., 61 (51), 67, 115 (50), 119

Kuno, H., 160, 164 (2), 160, 164 (3), 176

Kurath, S. F., 98 (25), 103

Kurihara, K., 160, 164 (2), 176

L

Lamb, P., 127 (30), 127 (31), 130

Lamoreaux, R. H., 42 (19), 45

Landel, R. F., 33 (4), 44, 89 (4), 101, 98 (24), 102

La Nieve, H. L., 116 (58), 120

Lee, B-L, 62, 63 (61), 68, 108 (30), 118

Legge, N. R., 63 (69), 68

Lenk, R. S., (5), 10, 125 (29), 130

Levett, C. T., 81, 82 (35), 85

Lewis, T. B., 139 (14), 155

Lidorikis, S., 111, 115 (35), 118, 128 (40), 131

Lindeman, L. R., 51 (20), 66, 72 (12), 84, 94 (20), 102

Litovitz, T. A., 42 (17), 45, 42 (18), 45

Lodge, A. S., 105, 106 (1), 116, 108 (31), 118, 124 (12), 124 (13), 129

Long, V. C., 81, 82 (32), 85

Loshaek, S., 69 (3), 83

Lovell, S. E., 90, 98 (6), 101

Lupton, J. M., 127 (32), 130

Lyons, J. W., (11), 10, 11, 12, 18, 22 (1), 27, 48 (2), 65

Lyons, P. F., 90, 92 (11), 102

M

McKelvey, J. M., (6), <u>10</u>

MacLeod, N. D., 153 (42), <u>157</u>

Macdonald, I. F., 106 (12), <u>117</u>

Macedo, P. B., 42 (17), 42 (18), <u>45</u>

Macosko, C., 20, 24 (13), <u>28</u>, 83 (49), <u>86</u>

Mackawa, E., 90 (7), <u>102</u>

Malkin, A. Ya., 61 (51), <u>67</u>, 63 (74), <u>68</u>, 115 (50), <u>119</u>

Markovitz, H., 24 (26), <u>28</u>, 96 (23), <u>102</u>, 106 (5), <u>116</u>, 106 (6), 106 (9), <u>117</u>

Maron, S. H., 14 (6), <u>27</u>

Marrucci, G., 124 (8), <u>129</u>

Marvin, R. S., 24 (30), <u>28</u>

Mason, J. H., 82 (44), <u>86</u>

Mason, S. G., 56 (38), 56, 57, 59 (40), 59, 60, 61 (45), <u>67</u>, 142 (16), <u>155</u>, 143 (23), 145 (28), <u>156</u>

Massati, F. G., 83 (49), <u>86</u>

Masuda, T., 54 (31), <u>66</u>, 69, 75 (6), 75, 80 (22), <u>84</u>, 80 (24), 80, 81, 82 (25), 81 (34), <u>85</u>, 96 (22), <u>102</u>, 108 (23), <u>118</u>, 148 (37), <u>156</u>

Matsumoto, T., 148 (37), <u>156</u>

Maxwell, B., 24 (33), <u>29</u>, 43 (25), <u>45</u>, 54 (32), <u>66</u>

Meier, D. J., 63 (71), <u>68</u>

Meissner, J., 23 (21), <u>28</u>, 109 (33), <u>118</u>, 124 (16), 124 (17), <u>130</u>

Mendelson, R. A., (7), <u>10</u>, 36, 38 (7), 36 (8), 36 (9), <u>44</u>, 81, 82 (29), 82 (37), <u>85</u>, 106, 115 (16), <u>117</u>, <u>115</u> (44), <u>119</u>

Merz, E. H., 12 (4), <u>27</u>, 54 (29), <u>66</u>, 114 (39), <u>118</u>

Metzner, A. B., 23 (22), <u>28</u>, 106 (7), 106, 115 (15), <u>117</u>, 108 (28), <u>118</u>, 122, 124 (2), <u>129</u>, 124 (19), <u>130</u>, <u>154</u> (48), <u>157</u>

Metzner, A. P., 23 (22), <u>28</u>, 122, 124 (2), <u>129</u>

Mewis, J., 124 (19), <u>130</u>

Middleman, S., (8), <u>10</u>

Miles, D. O., 24 (27), <u>28</u>

Milewski, J. V., 135 (5), 135 (6), <u>155</u>

Miltz, J., 81, 82 (30), 85
Minagawa, N., 115 (47), 119
Mooney, M., 57 (43), 67, 135, 153 (3), 155
Moore, W. R., 153, 154 (44), 153 (47), 157
Mori, Y., 106, 115 (17), 117
Moroni, A., 61 (52), 67, 72 (8), 84, 115 (51), 119
Morris, H. L., 62 (56), 67

                                  N

Nakagawa, T., 98 (26), 103
Nakajima, N., 115 (54), 119
Nason, H. K., 12 (3), 27
Nelson, P. A., 135 (8), 155
Newlin, T. E., 90, 98 (6), 101
Newman, S., 61 (53), 67, 115 (46), 119
Nicodema, L., 123, 124 (7), 129
Nicolais, L., 123, 124 (7), 129
Nielsen, L. E., 24 (25), 28, 49 (9), 49 (10), 65, 52 (26),
    66, 80 (26), 82 (40), 85, 82 (43), 86, 87 (1), 101, 114
    (39), 118, 135, 137, 153 (4), 135, 145, 146, 148 (7),
    137 (10), 139 (14), 155, 160, 162, 164, 171, 173 (5),
    176
Nissan, A. H., 106 (10), 117
Noel, F., 99 (31), 103
Noll, W., 106 (5), 116
Nyun, H., 43 (26), 45

                                  O

Ogihara, S., 44 (27), 45, 54 (30), 66
Ohta, Y., 80, 81, 82 (25), 85
Oka, S., 11, 12, 18, 22 (2), 27
Okagawa, A., 142 (16), 155
Olabisi, O., 108 (29), 118
Onogi, S., 54 (30) 54 (31), 66, 69, 75 (6), 75, 80 (22),
    84, 80 (24), 80, 81, 82 (25), 85, 148 (37), 156

P

Padawer, G. E., 142 (20), 155
Padden, F. J., 96 (23), 102
Park, J. Y., 122, 123, 124 (6), 129
Paul, D. R., 128 (38), 131
Pearson, G. P., 153, 154 (44), 157
Pearson, J. R. A. (9), 10, 125 (27), 130
Petersen, J. F., 22 (14), 28, 108 (26), 118
Pezzin, G., 90 (12), 90 (13), 102
Philippoff, W., 106 (11), 108 (18), 117
Pilpel, N., 162, 171 (13), 162 (14), 162 (15), 177
Plazek, D. J., 24 (29), 28
Pliskin, I., 115 (52), 115 (53), 119
Podolsky, Yu Ya, 115 (57), 119
Pollett, W. F. O., 20 (12), 28
Polyakov, O. G., 61 (51), 67, 115 (50), 119
Porter, R. S., 31 (2), 44, 40 (15), 45, 69, 72 (4), 74 (15), 84, 129 (42), 131
Prickard, J. H., 50 (15), 65

R

Rabinowitsch, B., 14 (15), 27
Radushkevich, B. V., 23 (20), 28
Ram, A., 81, 82 (30), 85, 153 (45) 157
Rankine, W. J. M., 161 (6), 176
Rautenbach, R., 22 (14), 28, 108 (26), 118, 160, 164 (4), 176
Regester, J. W., 127 (32), 130
Reisner, W., 162 (16), 177
Richards, J. C., 162, 164 (10), 177
Rieke, J. K., 82 (41), 86
Roberts, J. E., 106 (8), 117
Rodriguez, F., 61 (50), 67, 92 (16), 94 (18), 102, 115 (49), 119
Rogers, M. G., 114, 115 (41), 119

Roller, M. B., 83 (48), 86

Roovers, J. E. L., 81, 82 (31), 81 (34), 85, 96 (22), 102, 108 (23), 118

Rosen, S. L., 61 (50), 67, 115 (49), 119

Rudin, A., 74 (16), 84

Rudd, J. F., 74 (14), 84

Rumscheidt, F. D., 59, 60, 61 (45), 67

Rutgers, R., 135 (2), 155

S

Sabia, R., 74 (18), 84

Sailor, R. A., 106, 115 (15), 117

Sakoonkim, C., 98 (25), 103

Sanchez, I. C., 42 (20), 45

Sasaki, H., 44 (27), 45

Saunders, F. L., 151 (38), 156

Saunders, P. R., 90, 98 (6), 101

Schaller, E. J., 151 (41), 157

Schneider, Z., 153 (45), 157

Schonfeld, E., 82 (39), 85

Schott, H., 31 (1), 44, 52 (24), 66

Schowalter, W. R., 59 (46), 67

Schreiber, H. P., 72 (9), 84, 125, 128 (25), 130

Schümmer, P., 22 (14), 28, 108 (26), 118

Scott Blair, G. W., 148 (36), 156

Senna, M., 160, 164 (3), 176

Severs, E. T., (10), 10, 153 (43), 153 (46), 157

Shah, P. L., 99 (30), 99 (34), 103

Shaw, M. T., 99 (35), 103

Sherman, P., 56, 57, 59 (39), 67

Shida, M., 82 (38), 85, 115 (54), 119

Shih, C. K., 40 (14), 45, 74 (19), 84

Shroff, R. N., 54 (27), 66, 80 (28), 82 (38), 85

Sieglaff, C. L., 99 (29), 103

Simha, R., 56 (36), 66, 142 (17), 155

Simmons, J. M., 54 (34), 66, 98, 99 (27), 103
Simon, R. H. M., 40 (10), 44, 72, 74 (10), 84
Siskovic, N., 108 (32), 118
Slonaker, D. F., 115 (45), 119
Smith, R. W., 63 (66), 68
Soen, T., 61 (47), 61 (48), 67
Southern, J. H., 62 (57), 62 (60), 68, 128 (38), 131
Spencer, R. S., 125, 128 (22), 130
Spreafico, C., 61 (52), 67, 115 (51), 119
Spriggs, T. W. 106 (12), 117
Starita, J. M., 20, 24 (13), 28, 63 (65), 68
Stehle, R. G., 146 (32), 156
Stern, D. M., 98 (25), 103
Storey, S. H., 114 (40), 119
Stratton, R. A. 69, 71, 72 (5), 84, 106 (11), 117

T
Tadmor, Z., 108, 111 (27), 118
Taggart, W. P., 127 (35), 131
Takaki, T., 44 (27), 45
Takano, M., 160, 162, 164, 171, 173 (5), 176
Takashima, A., 148 (37), 156
Takayanagi, M., 24 (31), 29
Tanner, R. I., 108 (24), 118, 115 (55), 119
Taylor, G. I., 56 (41), 59 (44), 67
Taylor, J. S., 88 (3), 101
Tee, T-T., 54 (33), 66
Thimm, J. E., 18 (9), 27
Tobolsky, A. V., 63 (70), 68, 90, 92 (11), 102
Tokita, N., 115 (52), 119
Tokiura, S., 44 (27), 45
Tordella, J. P., 125, 127 (23), 125, 127 (24), 130, 128 (36), 131
Train, D., 164 (17), 177
Treloar, L. R. G., 49 (6), 49 (7), 65

Trementozzi, Q., 61 (53), 67, 115 (46), 119
Tsebrenko, M. V., 63 (67), 68
Turnbull, D., 42 (21), 45, 90 (14), 102

                                U

Utracki, L. A., 81, 82 (31), 85, 96 (21), 102, 111 (36),
    118, 128 (37), 131

                                V

Valentin, F. H. H., 162, 171 (9), 177
Vand, V., 139 (12), 155
van Krevelen, D. W., 35, 36 (6), 44
Vanoene, H., 63 (64), 68
van Selms, F. G., 146 (31), 156
Van Wazer, J. R., (11), 10, 11, 12, 18, 22 (1), 27,
    48 (2), 65
Verser, D. W., 54 (32), 66
Vinogradov, G. V., 23 (20), 28, 61 (51), 67, 63 (67),
    63 (74), 68, 115 (50), 115 (57), 119
Vlachopoulos, J., 111, 115 (35), 118, 128 (40), 131
Vranken, M. N., 24 (29), 28

                                W

Wagner, H. L., 75 (20), 84
Wagner, M. H., 124 (18), 130
Wales, J. L. S., 99 (36), 103
Wall, F. T., 49 (3), 65
Weemes, D. A., 115 (45), 119
Weissenberg, K., 20, 24 (11), 27, 105, 108 (2), 116
Weltman, R. N., 146 (34), 156
Werkman, R. T., 81, 82 (35), 85
West, D. C., 72 (9), 84, 114 (40), 119
Westover, R. E., 43 (24), 45
Westphol, S. P., 92, 96 (17), 102
Weymann, H. D., 145 (25), 156
White, J. L., 62, 63 (61), 68, 106, 115 (15), 117, 108
    (30), 118, 115 (47), 119, 124 (14), 130

Whitlock, M., 154 (48), 157
Williams, A. G., 116 (58), 120
Williams, M. C., 51 (21), 51 (22), 66, 92, 94 (15), 102, 108 (29), 118
Williams, M. L., 33 (4), 44, 80 (27), 85, 89 (4), 101
Wissbrun, K. F., 50 (15), 65, 75 (20), 84
Wohrer, L. C., 82 (44), 86
Wood, G. F., 106 (10), 117
Woods, M. E., 151 (39), 156
Wooley, R. H., 162 (8), 177
Work, J. L., 62 (58), 68
Wu, T. T., 142 (21), 155
Wyman, D. P., 40 (16), 45

                         Y

Yavorsky, P., 24 (26), 28
Yoshino, M., 24 (31), 29
Yudin, A. V., 63 (67), 68

                         Z

Zapas, L. J., 24 (26), 28, 96 (23), 102
Zidan, M., 124 (11), 129

SUBJECT INDEX

A

Aggregation, effect on viscosity, 139, 143
Andrade equation, 31
Angle of repose, 164
Apparent viscosity, definition, 5
Arrhenius equation, 31

B

Bagley correction, 16
Block polymer viscosity, 54, 61
Branching, 80, 115

C

Capillary viscometers, 12
Coaxial cylinder viscometers, 18, 160
Coherence of powders, 169
Complex viscosity, 7, 26, 52
Cone and plate rheometers, 20
Consistency, definition, 5
Coulomb equation, 170

D

Die swell, 106, 111
Dynamic mechanical properties, 52, 75, 96, 110, 114
Dynamic rheometers, 24
Dynamic shear modulus, 7, 26, 52, 75, 78, 96, 110
Dynamic viscosity, 7, 26, 52, 75, 78, 96, 110

E

Einstein coefficient, 56, 133, 134, 139, 142
Einstein equation, 56, 133
Elasticity, 2, 49, 75, 82, 96, 105, 111, 127
Electrostatic charge, effect on viscosity, 151
Elongational flow, 121, 127

E

Elongational viscosity, definition, 8, 121
Emulsions, 54, 56
Energy of activation, 31, 32, 34
Entanglements, **48**, 69, 80, 89
Entanglement molecular weight, 70
Extensional flow, 111, 121, 127
Extensional viscometers, 23

F

Filled polymers, die swell of, 115
Filled polymers, tensile viscosity of, 124
Filled polymers, viscosity of, 133
Flow instability, 125

G

Glass transition temperature, 87
Graft polymers, 61

I

Instruments, 11, 160
Instruments for powders, 160

K

Kelley-Bueche theory, 90

L

Latices, viscosity of, 137, 150
Logarithmic mixture rule, 63, 90
Lyons-Tobolsky theory, 92

201

L

Lubricants, 99

M

Master curves, 36, 72, 75
Melt fracture, 125
Molecular weight dependence
  of melt fracture, 128
Molecular weight dependence
  of viscosity, 69
Molecular weight distribution,
  52, 72
Molecular weight, dynamic
  properties, 75
Mooney equation 57, 135, 154

N

Newtonian viscosity, 3, 5,
  54, 121, 133
Non-Newtonian behavior,
  47, 74, 111, 142, 153
Normal stresses, 8, 22, 105,
  115

O

Oscillatory rheometers, 24

P

Packing fraction, 57, 135,
  136, 142
Parallel plate viscometers,
  22
Plastisol resins, 173
Plastisol rheology, 152
Poise, 3
Polyblends, viscosity, 54, 61
Powder rheology, 159
Powder rheology, instruments,
  160
Power law equation, 48, 52
Pressure, effect on viscosity,
  42

R

Rabinowitsch equation, 14, 16

R

Rankine equation, 162
Rate of shear dependence,
  theories, 51
Rate of shear effects,
  47, 51, 59, 71, 94, 110,
  145, 151
Rheology, definition, 1
Rheology of emulsions,
  54, 56
Rheology of plastisols,
  152
Rheology of powders, 159
Rheology of suspensions,
  133
Rheometers, 11

S

Shear cells, 162
Shear modulus, 7, 26, 54,
  75, 96, 110, 114
Shear rate dependence,
  theories, 51
Shear rate effects, 47, 51,
  59, 71, 94, 110, 145,
  151
Shear rate-temperature
  superposition, 36
Shear strain, definition,
  5
Shear stress, definition,
  5
Shift factor, 38
Solutions, dynamic proper-
  ties of, 96
Solutions, rheology of, 87
Superposition, 36
Suspensions, electrostatic
  charge, effect of, 151
Suspensions, non-Newtonian
  behavior, 142
Suspensions, surfactants,
  effect of, 148, 150, 152
Suspensions, viscosity of,
  133
Suspensions, water bridges,
  effect of, 146

T

Temperature dependence of
    viscosity, 31, 94
Tensile viscometers, 23
Tensile viscosity, definition,
    8, 121
Thermosetting polymers, 83

U

Units, 3

V

Viscometers, 11
Viscosity, definition, 5
Viscosity measurement, 3
Viscosity, molecular weight
    dependence, 69
Viscosity of block polymers,
    54, 61
Viscosity of emulsions, 54,
    56

iscosity of filled polymers,
    133
Viscosity of latices, 137,
    150
Viscosity of plastisols, 152
Viscosity of polyblends, 54,
    61
Viscosity of solutions, 87,
    89
Viscosity, temperature
    dependence of, 31, 94

W

Water bridges in suspensions,
    146
W-L-F equation, 33

Y

Yield point of suspensions,
    142